ENERGY AND CELLS

DIMENSIONS OF SCIENCE
Series Editor: Professor Jeff Thompson

ENERGY AND CELLS

Chris Gayford

Ph.D., M.Ed., B.Sc.
Lecturer in Education
University of Reading

MACMILLAN

First published 1986

Published by
MACMILLAN EDUCATION LTD
Houndmills, Basingstoke, Hampshire RG21 2XS
and London
Companies and representatives
throughout the world

Printed in Great Britain by
Camelot Press Ltd,
Southampton

British Library Cataloguing in Publication Data
Gayford, Chris
    Energy and cells.—(Dimensions of
    science series)
    1. Cell metabolism   2. Energy metabolism
    I. Title   II. Series
    574.87'61   QH634.5

ISBN 0-333-39621-9

To John-Paul, Libby and Claire

Series Standing Order

If you would like to receive future titles in this series as they are published, you can make use of our standing order facility. To place a standing order please contact your bookseller or, in case of difficulty, write to us at the address below with your name and address and the name of the series. Please state with which title you wish to begin your standing order. (If you live outside the United Kingdom we may not have the rights for your area, in which case we will forward your order to the publisher concerned.)

Customer Services Department, Macmillan Distribution Ltd
Houndmills, Basingstoke, Hampshire, RG21 2XS, England.

Contents

Series Editor's Preface

This book is one in a Series designed to illustrate and explore a range of ways in which scientific knowledge is generated, and techniques are developed and applied. The volumes in this Series will certainly satisfy the needs of students at 'A' level and in first-year higher-education courses, although there is no intention to bridge any apparent gap in the transfer from secondary to tertiary stages. Indeed, the notion that a scientific education is both continuous and continuing is implicit in the approach which the authors have taken.

Working from a base of 'common core' 'A'-level knowledge and principles, each book demonstrates how that knowledge and those principles can be extended in academic terms, and also how they are applied in a variety of contexts which give relevance to the study of the subject. The subject matter is developed both in depth (in intellectual terms) and in breadth (in relevance). A significant feature is the way in which each text makes explicit some aspect of the fundamental processes of science, or shows science, and scientists, 'in action'. In some cases this is made clear by highlighting the methods used by scientists in, for example, employing a systematic approach to the collection of information, or the setting up of an experiment. In other cases the treatment traces a series of related steps in the scientific process, such as investigation, hypothesising, evaluating and problem-solving. The fact that there are many dimensions to the creation of knowledge and to its application by scientists and technologists is the title and consistent theme of all the books in the Series.

The authors are all authorities in the fields in which they have written, and share a common interest in the enjoyment of their work in science. We feel sure that something of that satisfaction will be imparted to their readers in the continuing study of the subject.

Preface

CELLULAR ENERGETICS AND EXCHANGES

The aim of this book is to provide an up-to-date discussion of some of the basic ideas of cell biology and biochemistry appropriate for students who are completing their advanced level studies and who are approaching undergraduate and other comparable courses. A problem faced by many students at this stage is that explanations of some of the basic biochemical and physiological phenomena that are acceptable at GCE A-level, stop significantly short of what is expected in more advanced texts. This book sets out to provide a useful bridge between these two levels.

The topics included are those that are often considered to be difficult by students and teachers alike. Part of the purpose of the book is to reconcile what students may be taught in physical science courses with the biological topics included here, and in this way to provide an intellectually satisfying and coherent view of the subject. Also, it has been the intention to convey something of the way in which science is a growing and developing area of human activity with its own method of enquiry and creative speculation.

Throughout the text an effort has been made to use units and chemical nomenclature accepted by most authorities. Occasionally this has been problematic because some of the older terms are so well established in the literature that a change of name at this stage is likely to cause confusion. Such organic compounds are pyruvate whose new name is 2-oxopropanate, and ketoglutarate which is used in preference to 2-oxoglutarate.

Energy — An Introduction

The problem when writing about energy is that there is no satisfactory definition to begin with. Most physics texts start discussions about energy with definitions of work; this can have some helpful things to say to biologists, since work relates to change of energy in a system. In this way there is a concentration on the transfer of energy; however, this is not particularly helpful in thermodynamics but it does focus attention on energy conversion. The idea of the relationship between energy and work is only really helpful in mechanics, and it is a pre-condition in some other processes such as lighting and heating.

Broadly, energy can be said to be the ability of material systems to bring about changes in themselves or the environment. Even this type of statement causes difficulties since it depends on what is meant by 'change'.

The concept of energy shows one of the important characteristics of science in that it is an abstract idea, which although very useful is still only an aid to understanding. When thinking about energy there is a tendency to think of it as a commodity, but really it probably relates to a potential. Thus it is more helpful to think of ways in which energy can be obtained from a system rather than of a body possessing energy.

Really, no simple statements summarising energy are adequate. However, energy is absolutely fundamental to cell physiology and some understanding of its nature and interrelationships in the functioning of cells is essential. Perhaps we can only begin to understand energy through the network of concepts and laws which make up its supporting theories. These relate to the conversion, exchange and conservation of energy and the two laws of thermodynamics. It is the purpose of this short book to throw some light on these associated concepts in a biological context so that students may be able to better understand the relationships that exist between energy and living organisms and the total dependence that life has on energy. Also it is an area of science where impressive discoveries have been made and where researchers have used remarkable ingenuity in order to piece together even the partial picture that we have today.

1 Energy

Energy is of central importance to all living things. There are many activities and processes that sustain life and take place inside the cells of living things and energy is needed for them to take place. Such activities include synthesis of a large range of organic compounds, active uptake into cells of ions against concentration gradients between the cell and the environment, movement of part or the whole of an organism, and conduction of nerve impulses (figure 1.1).

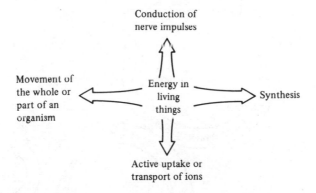

Figure 1.1

While the range of energy-demanding activities is large, many of the basic principles relating to energy are the same and the purpose of the first part of this book is to explore some of these principles.

Energy in the environment is often in a form that cannot be used by organisms. However, living things have the ability to convert unusable energy into forms that can be used in their cells and this is a fundamental requirement for life.

Thus in biology, energy is involved in processes where it is converted from one form into another. These energy-conversion processes are therefore very important and they are, in fact, fundamental to life. Photosynthesis and respiration are the two conversion processes and they supply energy in a form which can be used to drive the great range of activities in cells. Photosynthesis is the means by which an outside source of energy in the form of light or solar energy is converted to chemical energy in the cells of living plants. This provides the basis of the energy as well as the materials of all living things. Respiration begins with the food and energy made available by photosynthesis and converts this energy into a form that is readily usable and available within living cells (figure 1.2).

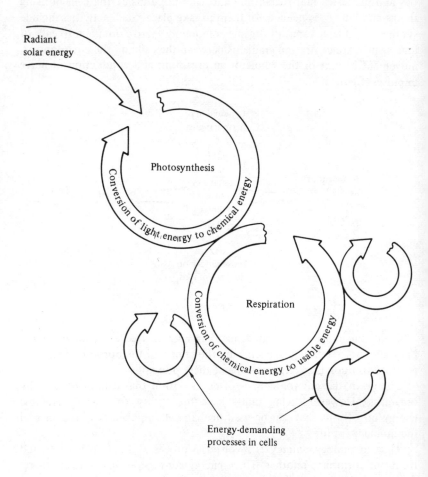

Figure 1.2

WHAT IS ENERGY?

Before we discuss the energy-conversion processes in biology, it may be helpful at first to consider some other aspects of the nature of energy and to mention some of its properties. Already some important things have been said about energy in the introduction, but further discussion is required.

Energy exists in a number of different forms such as light energy, mechanical energy and heat (figure 1.3).

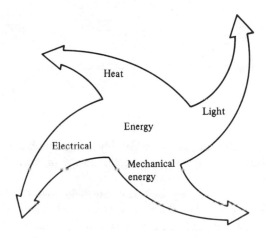

Figure 1.3

Simply speaking, energy would be defined by a physicist as 'the ability to do work'! For our purposes this statement is not very helpful and we need to delve more deeply into what is actually going on inside matter. First let us consider what is happening at the molecular level.

The atoms and molecules themselves show a number of different types of movement. This movement is often referred to as **kinetic energy**. For instance, molecules are moving about from one place to another in a random way, sometimes colliding with other molecules. This form of kinetic movement or energy is called **energy of translation**.

Also the electrons of the atoms in the molecules are moving by orbiting the nucleus of the atom; thus each electron of each atom has kinetic energy of its own. Not only are these forms of movement going on but also the atoms in the molecules are constantly vibrating relative to each other and the whole molecule is rotating as well (figure 1.4). There are thus three main types of movement involved. These atomic and molecular

3

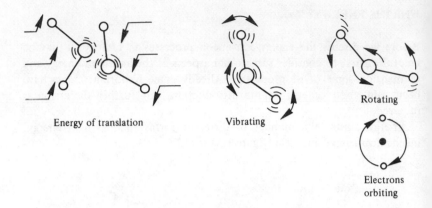

Energy of translation   Vibrating   Rotating

Electrons orbiting

Figure 1.4 *Different types of kinetic energy.*

movements are greatest in gases, considerably less in liquids and even further reduced in solids. As temperature increases, so the molecular and atomic movements also increase and speed up.

The atoms which make up molecules are held together by forces of electrical attraction between the positively charged central nuclei of the atoms and the negatively charged electrons orbiting around it.

There are also forces of repulsion that exist within the same molecule between similar charged particles. These forces of attraction and repulsion are exactly balanced and they are therefore in equilibrium. Living things are able to accumulate essential chemicals within their cells. These chemicals are a form of stored energy which may be called **potential chemical energy**. The term 'potential energy' is often used. This means that when this equilibrium between the different forces is disturbed in a chemical reaction, energy is released or absorbed as a result of changes in the equilibrium. Thus when chemical bonds are broken or created, changes in the equilibrium occur which result in release or absorption of the energy within the cell.

The different forms of energy, such as light, mechanical energy and heat, are all the result of kinetic energy of the atoms and molecules. The calculation of the sum total of the energy, both potential and kinetic, of just a single molecule is an extremely difficult task, and even when it has been achieved it is not very useful. In the study of biochemistry we are much more concerned with the energy changes that go on when reactions take place rather than with absolute energy levels.

THERMODYNAMICS

This is rather an off-putting term and in the narrow sense it is used to mean the study of the relationship between heat and other forms of energy. However, in a more general way it relates to energy in all of its forms and it does have some useful things to say to biologists who want to understand the fundamentals of energy and its relationship to living things.

A major principle of thermodynamics is that one form of energy can be converted into another. This is an important idea in the study of living organisms because energy conversions are an essential process, as has already been mentioned.

The principles of thermodynamics further state that when energy conversions occur there is no net loss or gain of energy, but that in all such processes, at least some of the energy is converted into heat which then tends to be lost. A simple non-biological example of these principles is an electric light bulb. The bulb is an energy converter in which electrical energy is converted mainly to light, but some of the energy is converted into heat instead. The heat is lost to the surrounding environment (figure 1.5).

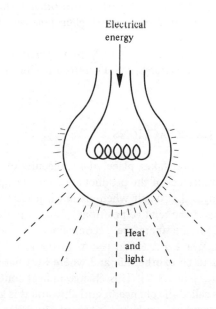

Electrical
energy

Heat
and
light

Figure 1.5

5

What we have considered so far can be summarised in the first and second laws of thermodynamics. The **first law** states that energy cannot be either created or destroyed. That is to say that there is no net loss or gain of energy in these reactions and also that the energy that you start with is equal to the amount that results from the reaction. The **second law** states that whenever an energy-conversion process takes place, then at least some of the energy takes the form of heat.

The scientist Joule developed a system of units which is used in thermodynamics and it can be used as a measure of all forms of energy. This means that these units can be usefully employed in energy conversion. The unit is the Joule and it is now the universal unit of energy, although in many texts you will often see calories quoted instead of Joules, especially in relation to diet and medicine (1 calorie is equal to about 4.2 Joules). The Joule is a very small unit and values often result from reactions which are measured in many thousands of Joules; thus the kiloJoule is often used (1 kJ = 1000 Joules).

As we have seen previously, the energy of organic molecules or compounds which is useful to living organisms is considered to be 'chemical energy'. This refers to the energy which is within the actual chemical bonds which hold the atoms together in the molecule. Compounds possess this chemical energy, which can be looked upon as potential energy, and this is released only when it is converted into other forms of energy, for instance heat. This bond energy has higher positive values for stable covalent bonds.

Heat is a special form of energy which, to all practical purposes, cannot pass from a colder to a hotter body. Therefore it will move in one direction only.

TYPES OF CHEMICAL REACTIONS

When a chemical reaction takes place, the molecules of the reactant are disrupted and the molecules of the product are formed (figure 1.6).

The energy changes that occur when chemical reactions take place are the result of changes in the relative positions of atoms in the molecules.

Chemical reactions are therefore accompanied by the production or absorption of heat. Where a reaction results in the loss of heat to the surroundings it is said to be **exothermic** and where heat has to be taken in it is then **endothermic** (figure 1.7). This change in heat content taking place during a reaction is called the change in **enthalpy** and it is given the symbol ΔH. When heat is given off, enthalpy is negative $(-\Delta H)$ and energy is lost as heat. Most reactions that proceed spontaneously produce heat and

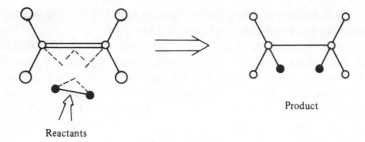

Figure 1.6

therefore have negative enthalpy. Endothermic reactions have positive enthalpy $(+\Delta H)$. Thus enthalpy tells us something about the energy conversions that are going on. Absolute enthalpy (H) is a measure of the internal energy of a system; thus it is the total of the many energy constituents of that system. It is not a very helpful value even if it could be measured satisfactorily. All energy changes show heat changes in a reaction. Thus all reactions are characterised by heat changes. This is because energy changes are involved in all reactions and some of these energy changes take the form of heat, as was made clear when we introduced the second law of thermodynamics.

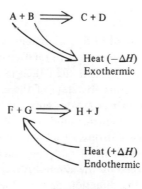

Figure 1.7

SYSTEMS

When discussing thermodynamics, scientists often refer to the situation within a 'system'. The term 'system' relates to the energy and materials

7

within a particular entity, or whatever is being talked about. For biologists a system may refer to the situation within a living cell.

In thermodynamics there are three main types of system. One is called an **isolated system**, where there are no exchanges of energy or materials with the outside. The second is a **closed system**. In closed systems, although they are delimited from the surroundings, they are able to exchange energy with the outside. The third type of system is the **open system**, and here both energy and materials are exchanged with the outside. This last type of system is the one that applies to biology (figure 1.8).

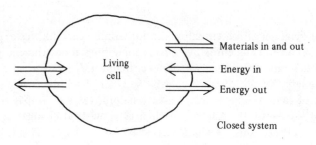

Figure 1.8

ENTROPY

Another term frequently used in thermodynamics in discussing the energy state of different systems is entropy (S). Often it is said that the more disorganised or chaotic the system then the greater is its entropy. This then is a term used to describe the state of the molecules. For instance, a compound in gaseous form has greater entropy than in liquid form, and both have greater entropy than the solid form. When a solid dissolves, so its entropy increases. The entropy of all compounds increases with temperature, because the movement of molecules increases with temperature. Entropy changes are denoted by the symbol ΔS.

Entropy also tells us the direction of spontaneous changes. If entropy increases $(+\Delta S)$ when a change occurs, then the change will be spontaneous. Since biological systems are open systems, the overall entropy change of the system and its surroundings must be considered. Obviously, since the environment is so immense, entropy changes in the surroundings are not measurable. Thus entropy change must be measured using indicators within the system, one of these being the flow of heat (enthalpy, ΔH) across the barriers of the system.

The relationship between entropy and enthalpy is that enthalpy represents the total amount of heat energy that accompanies a physical or chemical change. Not all of the energy is available or free to do work; some is retained by the system and increases the entropy. Thus ΔS is a measure of the energy which is not available to do work. Increases in entropy of a system mean that there is a decrease in energy availability. While entropy is a useful idea when trying to predict whether or not a reaction can occur, it is a phenomenon that is often difficult to measure. It involves changes not only in the system itself but also in the environment surrounding it. For this reason **free energy** rather than entropy is often used.

FREE-ENERGY CHANGE

Free energy is perhaps the most useful of all the thermodynamic concepts to biochemists. As you will see, free-energy change is produced by the combined changes in the enthalpy and entropy of a system. It is given the symbol ΔG, sometimes called the Gibbs free-energy change, and it is the form of energy capable of doing work. The relationship between free energy, enthalpy and entropy is often summarised in an equation $\Delta G = \Delta H - T\Delta S$ (for a change at constant temperature (T) and pressure).

Free energy is related to the energy potential of a molecule. This change in free energy that occurs when a reaction takes place is of particular significance.

In a reaction which yields energy, the energy may take a number of different forms, including, as has been mentioned previously, mechanical energy which will include volume changes, electrical energy, light and heat. When reactions occur under conditions which remain constant, then changes (such as pressure changes due to the exertion of mechanical energy through volume changes, or temperature changes caused by liberation of heat) will have to be dissipated from the system. Thus these forms of energy are not freely available to processes within the system. Other forms of energy, however, are generally 'free' to be used. The free energy that is yielded by a reaction can be looked on as the amount of energy that becomes available under constant conditions (figure 1.9).

When a reaction makes free energy available, ΔG is given a negative value. In reactions of this kind, the molecules that are reacting lose free energy when they become the products of that reaction. In such reactions, where free energy is lost from the molecules they are called **exergonic**; these reactions are usually spontaneous and they have a special significance in the energy relations of living things. You will notice that when discussing

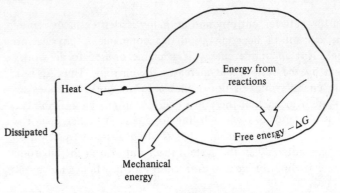

Figure 1.9

exothermic reactions these were also described as spontaneous. This is hardly surprising because reactions that release energy are also going to release some of that energy as heat.

When a reaction has a positive ΔG, then the reaction is referred to as **endergonic** and these reactions require an energy input from somewhere for them to occur at all (figure 1.10). As you will see later, in biological systems there are important endergonic reactions such as biosynthesis and muscle contraction, and these derive the energy needed by being linked to exergonic reactions within the same cell (figure 1.11).

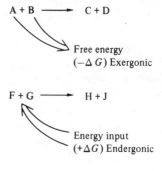

Figure 1.10

An example of a reaction that is exergonic overall is the oxidation of glucose in respiration. This may be simply written for our purpose here as

$$C_6H_{12}O_6 + 6O_2 \rightarrow 6CO_2 + 6H_2O + \text{energy}$$

The energy is represented here as a negative value where ΔG is given as

10

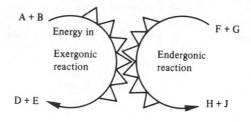

Figure 1.11 *Toothed wheel analogy.*

approximately 686 kJ per mole of glucose ($-\Delta G = 686$ kJ mol^{-1}). Moles are used because they are standard for all compounds and allow effective comparisons to be made. Also it should be noted that the expression of free-energy change for reactions is often given as the **standard free-energy change** for that reaction, written as $\Delta G^{\ominus}$, while the actual free-energy change in a cellular reaction is often written as ΔG. This standard free-energy change can be defined as the amount of free energy involved in the conversion of one mole of the reacting compound into the product under standard conditions; these standard conditions are taken as 25°C at normal atmospheric pressure and pH 7. If any of these conditions alter, then the free-energy yield will be altered as well. One of the problems is that standard conditions rarely occur in biological systems. Concentrations are usually far less than molar in cells; also the relative concentrations of substrate and product are very significant and the presence or absence of certain ions also strongly affects free-energy changes. As a result, exact measurement of all factors affecting the reaction is difficult and therefore agreement about free-energy changes for particular reactions is often difficult to achieve. Thus you will find different energy values given in different texts for the same reactions. Standard free-energy change values, although of some use for comparative purposes, tell us little about what is actually happening under the conditions that prevail in the cell. In fact for many reactions $\Delta G^{\ominus}$ may be a positive value but in the cell it may have a negative value. Also if the starting concentrations of the substrates for a reaction are relatively high and the products are continually removed so that they are kept low, then this has the effect of giving a reaction with a negative free-energy value (figure 1.12). Thus a reaction which is strongly exergonic can influence another which is endergonic by producing more substrate or removing the products. In this way one reaction affects the equilibrium of another, making it move more strongly in one direction or another. These are all highly significant factors affecting the direction of reactions inside cells. Processes involving all or some of these phenomena

11

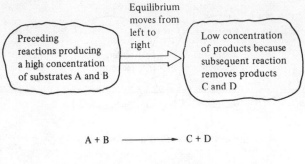

A + B ⟶ C + D

Figure 1.12

may alter what at first sight may not be considered a spontaneous reaction into one which is spontaneous. In this way such reactions are then able to take place inside cells.

BIOCHEMICAL PATHWAYS

Chemical reactions in living organisms usually take place in a series of sequence-linked reactions together, rather than as a single isolated reaction. The oxidation of glucose is a good example (see chapter 6). These sequences of reactions often make up complex biochemical pathways. A simple example to illustrate this principle may be

$$A + B \rightarrow C + D$$
$$\downarrow$$
$$D + E \rightarrow F + G$$
$$\downarrow$$
$$F + H \rightarrow J + K$$

The standard free-energy change for each particular reaction in the pathway may have either a positive or a negative value, but so that the sequence of reactions should proceed spontaneously there must be a negative value for the actual free-energy change for each step along the pathway. Thus, with reference to the pathway given above, the different stages may have the following changes in standard free-energy values; but all of the actual free-energy changes (ΔG) will have a negative value

$$A + B \rightarrow C + D \quad \Delta G^{\ominus} = -6 \text{ kJ}$$

$$D + E \rightarrow F + G \quad \Delta G^{\ominus} = +4 \text{ kJ}$$

$$F + H \rightarrow J + K \quad \Delta G^{\ominus} = -2 \text{ kJ}$$

It should be noted that when a reaction is exergonic $(-\Delta G)$ the value given is the maximum amount of energy that can be obtained from the reaction, and the efficiency of use of this will vary depending on the mechanisms used to couple this energy with other systems. Any energy that does not contribute to the ΔG value will usually be in the form of heat.

When chemical reactions occur, the chemical energy of the reactants is transferred to the products; but in exergonic reactions some of the energy is free to do work. Theoretically, the amount of work that can be done is energetically equivalent to the decrease in free energy. In fact, a chemical reaction can provide energy to perform work only if it can be harnessed in some way to utilise the energy (see chapter 8). Further discussion of chemical reactions and reaction rates will be found in chapter 4.

ENERGY – SUMMARY

1. There is no satisfactory definition for energy; but understanding some of the supporting theories and principles, such as energy conversion and conservation, is important.
2. Energy is required for many activities in living things, such as biosynthesis, active transport across membranes, movement and nerve-impulse conduction.
3. Photosynthesis and respiration are the two basic energy-conversion processes in living things. These provide energy in a form which can be used for energy-demanding functions.
4. Energy exists in the environment in different forms such as light, heat and mechanical energy.
5. Molecules, atoms and electrons are in constant movement, showing different types of kinetic energy.
6. Forces of electrical attraction and repulsion hold together the protons, neutrons and electrons of each atom.
7. Potential energy is the energy released or absorbed when chemical bonds are broken or made.
8. Different forms of energy, such as heat, light etc., are the result of kinetic energy.
9. One form of energy can be converted into another.
10. When energy conversion occurs there is no net loss or gain of energy, but some is converted into heat.

11. Energy changes take place when chemical reactions occur. These changes are related to alterations in the relative positions of atoms in the molecules.
12. Chemical reactions are accompanied by the production or absorption of heat. Reactions are therefore exothermic or endothermic. Changes in heat are called the enthalpy changes. Most spontaneous reactions have negative enthalpy.
13. All biological systems are thermodynamically closed, and they can exchange energy with the outside. The energy exchanged is normally heat or mechanical energy.
14. Entropy is used to describe the state of the molecules. The more disorganised they are, the greater the entropy.
15. If entropy increases, changes will be spontaneous.
16. Free energy (ΔG) is related to the chemical potential energy of a molecule, and it is concerned with the energy changes that occur when there is a reaction.
17. When a reaction makes free energy available, ΔG is negative and the reaction is spontaneous; such reactions are exergonic.
18. Endergonic ($+\Delta G$) reactions require an input of energy for them to occur. Therefore in order that endergonic reactions can occur in biological systems, they are linked to exergonic reactions.
19. The conditions that exist inside cells differ so markedly from standard conditions that a $+\Delta G$ reaction may even become a $-\Delta G$ reaction in the cell.
20. Biochemical reactions often occur in pathways. In this way the production of substrates from previous reactions and the removal of products subsequently can significantly alter the equilibrium of an individual reaction so that reactions with a positive free-energy change under standard conditions may proceed.

2 Adenosine Triphosphate

So far we have considered the nature of energy and the idea that life is sustained by a large number of interconnected sequences of reactions known as **metabolic** pathways. These pathways are of two main types, one resulting in synthesis and the other in breakdown of compounds.

Synthetic reactions result in the building up of more complex or elaborate compounds from simpler ones; examples are the production of proteins from amino acids and cellulose from glucose. Generally speaking these reactions are often involved in growth, repair or storage in organisms. Such reactions are called **anabolic** and they are usually endergonic ($\Delta G^{\ominus}$ has a positive value). Thus, an energy input is needed. These reactions result in an increase in atomic order and therefore a decrease in entropy (figure 2.1).

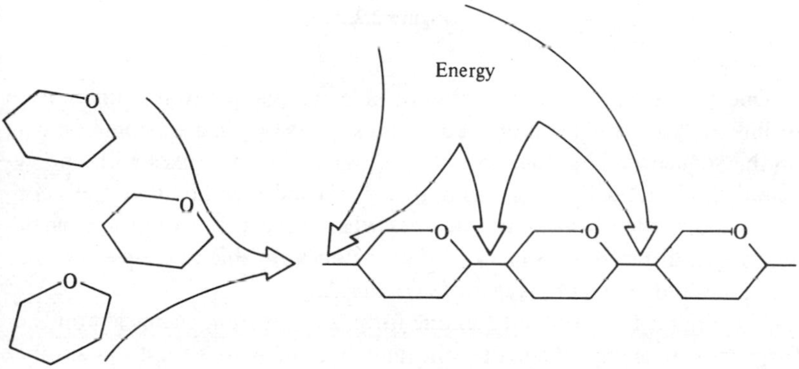

Figure 2.1

Breakdown reactions involve complex molecules which are divided up into smaller and simpler ones, as in the case of the respiratory pathway inside cells. These breakdown reactions are **catabolic** and usually involve the release of energy, consequently they are exergonic ($-\Delta G$). They result in a decrease in atomic order and there is a resultant increase in entropy (figure 2.2).

Even exergonic reactions rarely proceed spontaneously at a rate which is fast enough for living organisms. Most require a small input of energy and this is referred to as the **activation energy** (see chapter 4) whose purpose is to speed up the reaction; usually catalysts are involved and in biological systems these are enzymes. The effect of catalysts is to lower the amount of activation energy required, thus allowing the reaction to proceed more quickly.

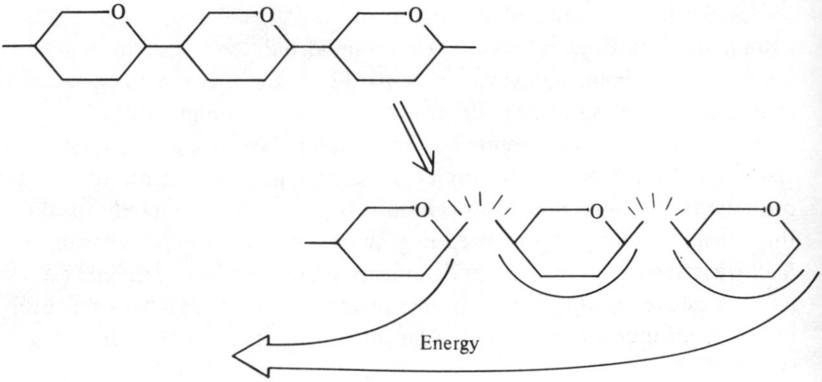

Figure 2.2

One of the ways that the network of metabolic pathways functions is to link anabolic and catabolic sequences so that they share a common part of the sequence. The coupling of energy-yielding processes with energy-demanding reactions is vital and in this way one drives the other. Cells contain a multitude of different enzymes where each is specific for a particular chemical reaction. Some of these enzymes are able to couple together exergonic and endergonic reactions (figure 2.3).

If energy is to be obtained in one form, such as from chemical bonds or from radiant energy from the sun, and it is then to be used in another form, such as the transport of molecules or to create new chemical bonds, there must be some way of linking the catabolic energy-releasing process with the anabolic energy-demanding process.

In living things a method for the temporary storage of the energy produced by exergonic processes is needed, so that it can be supplied to other endergonic reactions. In the absence of such a mechanism, the energy available from exergonic reactions would all be released as heat.

16

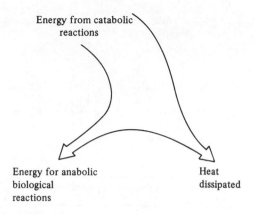

Energy from catabolic reactions

Energy for anabolic biological reactions

Heat dissipated

Figure 2.3

Heat has only limited use as a form of energy and it is quickly dissipated. Other more useful forms of energy are therefore needed. Of course, as has been explained in chapter 1, some of the energy will inevitably be released as heat, but it is the purpose of the biological systems under consideration here to keep this to a minimum.

There are two common types or groups of compounds which have an important function as intermediates in these energy-producing and utilising processes — these are **reduced co-enzymes** (see chapter 3) and high-energy phosphate compounds. The best known and most significant of these in cellular metabolism is **adenosine triphosphate** (ATP). A molecule of ATP consists of an adenosine section, which in turn is made up of adenine, which is a double-ring aromatic nitrogenous-based compound also found in DNA, and also a five-carbon sugar called ribose. However, from our point of view here the most important feature of the molecule is that the adenine is attached to three phosphate groups joined one to the other by a special type of chemical bond which is an **anhydride bond**, and they are then linked to the ribose section by an ester linkage. If instead of three phosphate groups there are only two, then the compound is adenosine diphosphate (ADP) — see figure 2.4. These anhydride bonds give particular properties to the ATP and ADP molecules.

Synthesis of ATP from ADP and inorganic phosphorus is the major way that energy from light (**photophosphorylation**, see chapter 7) and oxidation of foodstuffs (**respiratory phosphorylation**, see chapter 6) is conserved for use in the energy-demanding processes of the cell (figure 2.5).

An important feature of the ATP molecule is the presence of unstable 'energy-rich' anhydride bonds. The standard free energy ($\Delta G^{\ominus}$) of hydrol-

17

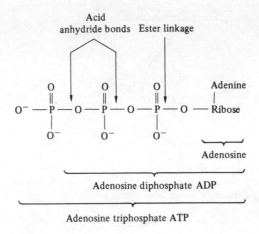

Figure 2.4 *Chemical structure of ATP and ADP.*

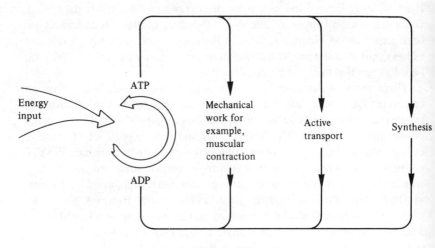

Figure 2.5

ysis of each of these bonds is about -30 kJ mol^{-1}. The linkages which attach the first phosphate group to the adenosine part of the molecule have a much smaller standard free energy of hydrolysis.

$$\text{ATP} + \text{H}_2\text{O} \xrightarrow[\text{enzyme}]{\text{ATPase}} \text{ADP} + \text{H}_3\text{PO}_4 \quad \Delta G^{\ominus} = -30 \text{ kJ mol}^{-1}$$

Even here it is necessary to be critical about this value of -30 kJ mol^{-1} for reasons which will be apparent by reconsidering chapter 1.

18

The value for ΔG for the anhydride linkage may vary considerably however, depending on the prevailing conditions that exist within the cell at the time. Some of the energy is lost as heat but a proportion can be used for biological activities.

When the energy of ATP is used for energy-demanding processes within the cell, the ATP is broken down to give ADP and inorganic phosphate (figure 2.6).

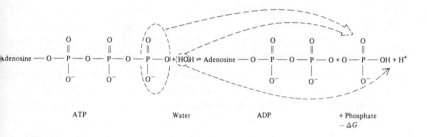

ATP   Water   ADP   + Phosphate
                                          $-\Delta G$

Figure 2.6

The term 'energy-rich bond' has been the cause of a great deal of confusion in the minds of biology students and it may therefore help if we briefly try to clarify the situation. The view that energy in biological systems is derived from the hydrolysis of ATP and that this energy is somehow contained in the terminal phosphate bond is unsatisfactory. Also that this energy is released when the bond is broken is not consistent with our understanding of the nature of chemical energy as outlined previously (chapter 1).

To a physical chemist **bond energy** refers to the energy associated with a particular chemical bond. The energy absorbed when the bond is broken, and released when the bond is formed, is called the **bond dissociation energy**. Thus the bond dissociation energy is the energy needed to break the covalent bond. This energy is high for stable bonds. Biochemists have not explained bond energy or bond dissociation energy in this way; here they have often related this to the difference in free energy of the products and free energy of the reactants.

Part of the problem is that it is not as though the terminal phosphate group of ATP is pulled from the molecule and nothing else happens. When the phosphate is removed it is replaced by water in the hydrolytic reaction. In this way, as in any chemical reaction, bonds are broken and bonds are formed. The overall change in free energy depends on the sum total of the free-energy changes for each part of the reaction. The symbol $\sim$ is often

19

used to indicate a 'high-energy bond'. This can be quite misleading since the free energy is not contained in the one bond, since this is a normal bond and it is not more highly 'energised'. Consequently the energy value for removal of this terminal phosphate is the standard free-energy change for the whole hydrolysis reaction. In this reaction the bond is broken and changes in the molecular structure occur. It is all of these changes in total which bring about the overall free-energy change. The energy therefore should be thought of as a characteristic of the entire molecule and the reaction in which it is involved, and not of a particular bond within the molecule. Also it should be understood that although the terminal phosphate of ATP may be represented by the symbol ~P, at hydrolysis when the phosphate is liberated it does not take the 'high-energy' with it but behaves like an ordinary inorganic phosphate group. The term 'high-energy' or 'energy-rich' compound means that the molecule, rather than the bond, has a characteristic large negative free energy of hydrolysis. Also, there is no distinction between 'energy-rich' and 'energy-poor' compounds. In fact, there exists in biological systems a complete range of compounds extending from those with a very large negative free energy of hydrolysis to those where this is small. Thus the term 'high-energy' or 'low-energy' is arbitrary. This phenomenon is significant because the energy-carrying properties of ATP depend on it. If you look at table 2.1 you will see a list of compounds in a decreasing series of free energy of hydrolysis. As an approximate guide it may be considered that compounds showing a standard free energy of hydrolysis which is more negative than -29 kJ at pH 7 may be thought of as 'energy-rich'.

Generally, any compound may be exergonically phosphorylated by taking on phosphate from any compound in the series with a more negative $\Delta G^{\ominus}$. In this way, if you look again at table 2.1, ADP can be phosphorylated to ATP by phosphate transfer from compounds such as phosphoenol pyruvate or phosphocreatine, but not glucose-6-phosphate.

It is significant that ATP occupies an intermediate position in the range of phosphorylated compounds (that is, those that have had phosphate added to them) in biological systems. The basis of the role of ATP is that in the

Table 2.1

	$\Delta G^{\ominus}$ (kJ mol^{-1})
Phosphoenol pyruvate	-54.6
Phosphocreatine	-42.0
ATP $\rightarrow$ ADP + P$_i$	-31.0
Glucose-6-phosphate	-13.9

ATP–ADP conversion process the molecule is able to act both as an acceptor and a donor of 'energy-rich' phosphate groups. What happens is that ATP accepts phosphate from even higher-energy phosphate compounds and is able to donate it to lower-energy phosphate compounds (figure 2.7).

One example of a reaction where ATP is formed is the coupling of the synthesis of ATP with the conversion of phosphoenol pyruvate (PEP), a compound produced in the glycolytic pathway (see chapter 6). In figure 2.7 you will see that the endergonic synthesis of ATP from ADP is coupled to the exergonic hydrolysis of PEP. This manoeuvre means that an enzyme-initiated reaction takes place spontaneously. It includes the synthesis of ATP which is endergonic and it still has an overall ΔG of about -23 kJ mol^{-1}. The phosphate group is donated by the PEP to ADP which then forms ATP. This is one of many coupled reactions where ATP synthesis occurs.

In the second stage of the function of ATP, where it acts as an energy carrier, it can donate its terminal phosphate group to a large number of different phosphate acceptor molecules. Among these is glucose, as shown also in figure 2.7. In this way the phosphate of ATP can be transferred to many different molecules which may themselves be used subsequently in

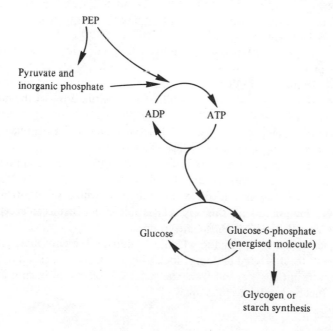

Figure 2.7

21

synthetic reactions. By this process the acceptor molecule, such as glucose, is 'energised' to glucose phosphate in preparation for the energy-demanding synthetic reaction that it will be involved in later (see also chapter 8).

There is a constant loss of energy from cells to the environment and therefore all living things need an energy supply from the environment which is continuous. This may be either by photosynthesis using radiant energy from the sun (as plants do) or by using the energy of food molecules taken in from the environment. Both of these forms of energy are converted into chemical energy (ATP) which can be used by the cell. However, the changes that food molecules, or the sugars synthesised in photosynthesis, need to undergo before they can be used to liberate energy are complex and will be dealt with later (chapters 7 and 8).

ATP – SUMMARY

1. Metabolic pathways are of two kinds – synthetic or anabolic ones which usually require an input of energy and breakdown, and catabolic ones which usually release energy.
2. In order that biological systems can work, anabolic and catabolic processes are linked so that one drives the other.
3. ATP is a chemical which is able to conserve energy from anabolic processes for use in catabolic ones.
4. The ATP molecule has unstable anhydride bonds which when hydrolysed produce a relatively high-energy change ($\Delta G = -30 \, \text{kJ mol}^{-1}$).
5. The free energy of ATP is not located in the phosphate bond but instead the free-energy change results from restructuring of the molecule when hydrolysis occurs.
6. High-energy compounds, of which there are many, all have a relatively large negative free energy of hydrolysis.
7. Compounds may be exergonically phosphorylated by compounds with an even more negative ΔG of hydrolysis.
8. ATP occupies an intermediate position in a whole series of 'high-energy' compounds. In this way ATP is able to act first as an acceptor and then as a donor of phosphate groups.
9. Frequently the compound to which ATP donates the phosphate group becomes the activated building block of synthetic reactions.
10. Energy is initially derived from the sun, as in plants, or from synthesised fuel molecules, as in animals.

3 Redox Reactions and Cellular Energetics

Energy, as has been discussed in chapter 1, exists in a number of different forms. Movement is associated with whole molecules, single atoms and each electron. Here we are particularly concerned with the energy associated with the movement of electrons around the inner core of protons and neutrons of atoms. In this chapter the central part played by the energy of the electrons in the energy relationships within cells will be discussed.

Most of the energy of the majority of living organisms is derived from the oxidation of hydrogen by oxygen, with the ultimate formation of water. In animals, this hydrogen is initially taken into the body as part of the chemical matter which makes up the molecules of food. In plants, the food molecules containing hydrogen are synthesised by the process of photosynthesis (chapter 7). The most common basic material for the beginning of energy in both plants and animals is glucose (see chapter 6, where a more detailed discussion of this part of the process can be found). However, at this stage it is important to realise that very large amounts of energy are evolved from the oxidation of molecules of glucose — too much to be of value to the organism if it is released in one simple reaction. Perhaps it is worth your considering why large amounts of energy produced in this way would not be of value to an organism. Instead a process has been developed in living things which allows small quantities of free energy to be released in a controlled way. This energy can be used for the synthesis of ATP from ADP.

OXIDATION AND REDUCTION

The processes of oxidation are important in cellular metabolism. Chemically, **oxidation** and the converse reaction, **reduction**, are two aspects of the same overall process. For a substance to be oxidised in a reaction it must first be in a reduced state. Reactions involving *red*uction and *ox*idation are often referred to as **redox** reactions. Chemically redox reactions are not only confined to those that involve the addition and removal of oxygen but they also include reactions where either **hydrogen** or **electrons** are added or removed. The overall effect is that the substance oxidised always loses electrons and the substance reduced always gains electrons. In this way each of the following types of reaction is considered to be a redox reaction

1. $\widetilde{X} + RO_2 \rightleftharpoons \widetilde{XO_2} + R$ (X is oxidised by the addition of oxygen and R is reduced)

2. $\widetilde{XH_2} + Z \rightleftharpoons \widetilde{X} + ZH_2$ (X is oxidised by the removal of hydrogen and Z is reduced)

3. $Fe^{2+} \rightleftharpoons Fe^{3+} + e^-$ (Fe^{2+} iron is oxidised to Fe^{3+} iron by the removal of an electron, and the addition of an electron would result in reduction)

In reactions of the first two types, when one substance is oxidised then the other is reduced and vice versa. The third type of reaction is a little different and it will be considered more fully later. Redox reactions are also reversible, so that if the reaction proceeds in one direction then one of the reactants is oxidised, and if it moves in the opposite direction then the same reactant is reduced. In practice what actually happens is that for an oxidation reaction to take place a corresponding reduction reaction must occur at the same time. If we consider the second type of reaction given above as an example of this principle you will see that when X is oxidised by the removal of hydrogen, Z becomes simultaneously reduced by the addition of hydrogen (figure 3.1). With these reversible redox reactions, the reduced and oxidised forms of the compound are referred to as a 'redox couple'. Consequently, any oxidation–reduction reaction involves two redox couples – for example, XH_2/X and Z/ZH_2.

Biological systems use mainly the second and third of the three types of redox reaction given above. That is, they usually involve hydrogen or elec-

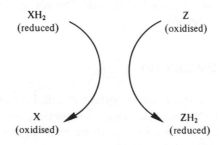

Figure 3.1

trons. For two redox couples to function together in these reactions there must be a difference in their affinity for either hydrogen or electrons. The redox reagent with the greater electron affinity is regarded as the oxidising agent (oxidant in the reaction); in this way it takes on the electrons or hydrogen, and becomes reduced. The donor is then said to be oxidised. In biological systems the role of redox couples is taken on by a range of organic compounds that are often protein based; these are co-enzymes and enzymes, and they have specific portions of the molecule that are capable of being reversibly oxidised and reduced. (For a further explanation of the nature and functioning of enzymes see chapter 4.) Co-enzymes are organic substances that work in association with enzymes and render those enzymes active.

To understand something of the thermodynamic processes going on, it should be realised that the electrons are negatively charged and that they are in an electric field associated with each atom. In order to move an electron further from the positive charge of the proteins and neutrons at the centre, work has to be done; that is, energy has to be expended ($+\Delta G$). When this happens there is an increase in the 'potential' energy. If, alternatively, the electron moves spontaneously towards the positive part of the atom then energy is lost ($-\Delta G$); this lost energy is free energy which is available for other purposes. Also when electrons are transferred from one substance to another, free energy may be released ($-\Delta G$); these reactions will be spontaneous (see chapter 1). The oxidation of certain co-enzymes is accompanied by the release of a large and useful amount of free energy (figure 3.2).

If enough energy is supplied to the system, then an electron may completely leave the electric field of an atom. The atom is then said to be oxidised. Very little energy is required to remove an electron from the

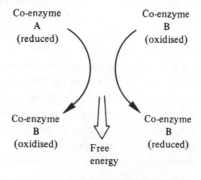

Figure 3.2

field of a hydrogen atom. Also, free hydrogen atoms themselves are extremely reactive and their free existence is of very limited duration. So when hydrogen atoms are removed from molecules, as they are in respiratory metabolism (see chapter 4), they either quickly become attached to another molecule or they dissociate to form hydrogen ions and separate electrons.

$$H \rightleftharpoons H^+ + e^-$$

The electron then rapidly becomes associated with another atom in a molecule. When this new association between the electron and another atom takes place there is a yield of energy $(-\Delta G)$.

The ease with which a particular reductant tends to lose electrons, or an oxidant gains electrons, is a measure of the **standard oxidation–reduction potential** $(E^{\ominus})$, often referred to as the **redox potential** of that substance. This value is measured under standard conditions of temperature, pH and in the presence of a molar concentration of the oxidant. Any redox reaction theoretically can be made to generate an electric current, indeed redox reactions are used to provide currents in batteries. In biological systems the current will be measured in milliamps. A negative value $(-E^{\ominus})$ or a positive value $(+E^{\ominus})$ indicates the direction of movement of electrons from one redox pair to another. The more positive the $E^{\ominus}$ value then the more powerful it is as an oxidising agent, and conversely the more negative the value of $E^{\ominus}$ the more powerful it is as a reducing agent. The tendency for substances to undergo redox reactions will differ from one substance to another and depends on the relative ease with which it gains or loses electrons. For comparative purposes this is measured against the ease with which hydrogen gas is oxidised to hydrogen ions. Those substances with a negative $E^{\ominus}$ value lose electrons more readily than hydrogen gas and those with a positive $E^{\ominus}$ value are less easily oxidised. It is possible to measure redox potentials and to plot the progress of redox reactions, even in living tissues, using microelectrodes. This can be useful in the determination of the energy economy and relationships of cells. In thermodynamics there is a relationship between $E^{\ominus}$ and $\Delta G^{\ominus}$. Consequently, from this relationship the free-energy change of a reaction can be calculated from measurements of the redox potential, although this will not concern us here.

Enough should have been said so far to show that redox reactions are also basically competitive and that whether the same substance acts as an oxidant or a reductant depends on what other substances are to be found in association with it and what their redox potentials may be.

Thus in a system such as the following

$$\text{AH}_2 \searrow \text{B} \searrow \text{CH}_2$$
$$\text{A} \swarrow \text{BH}_2 \text{C}$$

B is acting as an oxidant in relation to A but as a reductant in relation to C. A is a more powerful reductant than B and B is more powerful than C. This determines the direction of movement of hydrogen or electrons in the system, and by this mechanism a chain of reactions can take place.

At the beginning of this chapter it was said that the energy relationships of living cells involve the processes of hydrogen and electron transport. This takes the form of transfer of hydrogen or electrons from one enzyme or co-enzyme to another in a stepwise manner in a chain, making a sequence of several stages. These substances are often referred to as hydrogen or electron carriers. The co-enzyme changes its state from the reduced to the oxidised form and this reaction is linked to synthesis of ATP from ADP. The entire process is called **oxidative phosphorylation** because ADP is phosphorylated to ATP and the co-enzymes are changed from the reduced to the oxidised state.

The transfer of electrons from the reduced co-enzymes releases the free energy that is then used for ATP formation (figure 3.3).

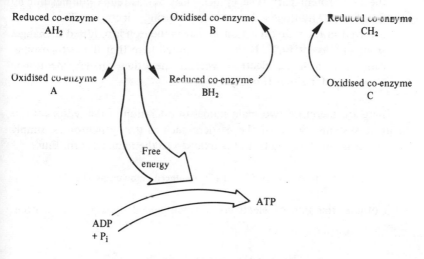

Figure 3.3

There are four main groups of redox couples in cells — these involve the following compounds.

(a) Nicotinamide adenine dinucleotide (NAD^+) and nicotinamide adenine dinucleotide phosphate ($NADP^+$) are co-enzymes associated with, but not forming part of, the molecule of dehydrogenase enzymes which pass hydrogen from the substrate to the NAD^+ or $NADP^+$ to become NADH or NADPH.

(b) Flavin may occur in one of two particular forms: either flavin mononucleotide (FMN) or flavin adenine dinucleotide (FAD). Both types are attached to much longer protein molecules and together these form flavoproteins. Thus the flavins can be said to be prosthetic groups for the enzyme complex (see Enzymes, chapter 4). In the reduced form they become $FADH_2$ or $FMNH_2$

(c) Co-enzyme Q (CoQ) or ubiquinone (UQ) consists of a ring-shaped part to the molecule which is a quinone compound. This is attached to a long-chain fatty acid. Quinones are easily reduced to form hydroquinone. This may be written as $CoQH_2$ or UQH_2.

(d) Cytochromes are a group of compounds which consist of a protein part and generally a non-protein portion that usually contains iron, rather like the iron-containing haem group associated with the protein part of a haemoglobin molecule in blood. The main cytochromes are simplified here and they are called cytochrome b, c, a and a_3. The last is sometimes called cytochrome oxidase and it also contains copper in the non-protein part. The cytochromes act as redox couples not by transferring hydrogen but by transferring electrons. The iron is involved in the redox reaction so that as the iron is oxidised it changes from $Fe^{2+} \rightleftharpoons Fe^{3+}$. It should be noted here that the cytochromes carry only a single electron whereas the other carriers mentioned above transfer either H or H_2.

There are therefore two main modes of operation of the redox couples in living systems. Some of the carriers, such as the cytochromes, simply gain electrons in changing from the oxidised to the reduced form thus

$$\text{oxidised cytochrome } (Fe^{3+}) + 1e^- \rightarrow \text{reduced cytochrome } (Fe^{2+})$$

With others, the gain of electrons is accompanied by a gain of protons, thus

$$XH_2 + NAD^+ \rightarrow X + NADH + H^+$$

or

$$\text{ubiquinone (UQ)} + 2e^- + 2H^+ \rightarrow UQH_2$$

Here it should be noted that either one or two protons are involved, depending on the carrier.

If we now briefly consider the system of carriers in respiration, some further principles should become more clear. These carriers form a final common pathway in the energy-conversion process of the cell, and along this all of the electrons from the fuel materials, such as glucose, used in respiration pass ultimately to oxygen in aerobic cells. The electrons entering the chain have a relatively high energy level and as they move along the chain a good deal of their energy is conserved in the formation of ATP. Basically, the pathway begins with NAD and then passes to a flavoprotein; often FAD is followed by ubiquinone or co-enzyme Q (figure 3.4a). After this there is a sequence of cytochromes finishing with cytochrome a_3 (cytochrome oxidase). This last stage in the chain links protons and electrons with oxygen to form water (figure 3.4b). You may wish at this point to consider what type of simple evidence an investigator would use to decide where a newly discovered substance may fit into an electron carrier sequence.

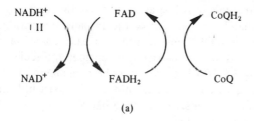

(a)

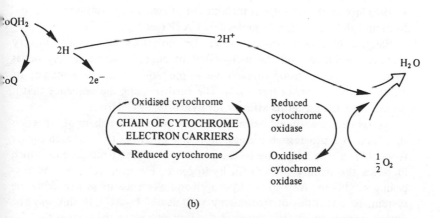

(b)

Figure 3.4

29

It will be seen that in the first part of the system there is a complex balance between protons and electrons related to the carrying capacity of the different co-enzymes involved.

In the second part of the hydrogen/electron transfer system, the hydrogen dissociates into protons and electrons. The protons go into solution as hydrogen ions and the electrons are passed along the system of carriers to be recombined with the protons at the end.

If we now relate what has been said so far to what was discussed in the previous chapter concerning free energy and ATP, it will be seen that the transport of hydrogen or electrons along the chain of redox couples is a sequence of exergonic reactions which yield free energy. The synthesis of ATP from ADP on the other hand is endergonic and these two types of reaction are coupled together. The series of redox reactions produces about -217 kJ mol^{-1} and the synthesis of ATP requires about $+30$ kJ mol^{-1}. Thus each chain of redox co-enzymes is theoretically capable of producing enough energy for the synthesis of several molecules of ATP. However, it must be remembered that according to the principles of thermodynamics, when energy-conversion processes such as these take place, then some of the energy is lost as heat. Accordingly, it can be calculated that approximately 40 per cent of the available energy is actually used in the synthesis of ATP. Consequently, for each complete passage of hydrogen or electrons along a carrier system, three molecules of ATP are produced, utilising 3×30 kJ. It also must be remembered that, as we have said before, standard conditions do not usually exist in living cells and therefore the actual free-energy levels are likely to vary considerably from the theoretical free-energy levels.

The amount of free energy released at each step in the chain of carriers varies. Only at certain steps is the amount of free energy released sufficient to permit the synthesis of a molecule of ATP (see figure 3.5).

Electron/hydrogen-transporting carrier molecules are arranged in very particular sequences on the membranes of energy-transducing organelles (chapter 5). The position of each one in the sequence is determined by its affinity for electrons or hydrogen. The further along the sequence that it occurs, then the higher its affinity for hydrogen or electrons.

As has been mentioned before, a characteristic of the chain of carriers is that there is a progressive increase in redox potential ($E^{\ominus}$) of each carrier from NAD at the beginning to the terminal oxidase at the far end, which increases the affinity of each for hydrogen or electrons relative to its preceding neighbour. Also, as the hydrogen and electrons are passed along the system, so quantities of free energy are released ($-\Delta G$); in this way the overall free energy available increases as you pass along the system.

30

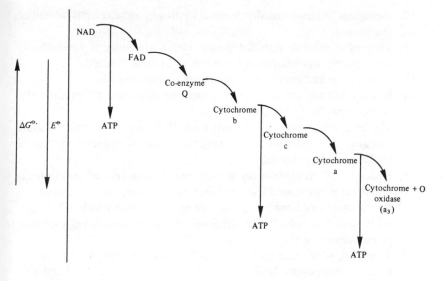

Figure 3.5

The other biochemical processes whereby energy production occurs through respiration in living cells will be considered in chapter 6.

REDOX REACTIONS – SUMMARY

1. Energy is associated with the movement of electrons.
2. Most energy for the majority of organisms comes from the oxidation of hydrogen by oxygen.
3. Hydrogen is taken into organisms as part of food molecules.
4. Energy is released in a controlled way from food molecules and this is used to synthesise ATP.
5. Oxidation and reduction reactions are important in energy metabolism and they are aspects of the same reaction.
6. Redox reactions may be of three types – addition or removal of either oxygen or hydrogen or electrons.
7. In cases where oxygen, hydrogen or electrons are involved, when one substance is oxidised another is reduced.
8. Normally when an oxidation reaction occurs, a corresponding reduction reaction occurs.
9. With reversible redox reactions, redox couples are involved.

10. Biological systems usually involve hydrogen or electrons in redox reactions.
11. The redox reagent with the greater electron affinity is the oxidising agent taking on electrons or hydrogen to become reduced.
12. Co-enzymes and enzymes are the redox couples in cells.
13. When electrons are transferred from one substance to another, the free-energy change is negative.
14. The ease with which a reductant loses electrons, or an oxidant gains electrons, is a measure of the standard oxidation–reduction potential ($E^{\ominus}$) or redox potential.
15. Redox reactions often occur as chain reactions in cells where electrons or hydrogen are passed from one enzyme or co-enzyme to another.
16. Redox reactions have a negative free-energy change which is linked to ATP synthesis, which is endergonic; the process is called oxidative phosphorylation.
17. There are four main groups of redox couples, which involve the following compounds: NAD and NADP, FMN and FAD, CoQ and UQ, and cytochromes.
18. The cytochromes transfer electrons.
19. For every complete passage of hydrogen or electrons along the carrier system, three molecules of ATP are produced. ATP is produced only at certain stages in the chain of carriers.
20. Carrier molecules are arranged in special sequences on energy-transducing membranes.
21. From NAD at the beginning of the chain to the terminal oxidase there is a progressive increase in redox potential.

4 Enzymes and Reactions

A large number of chemical reactions are spontaneous and occur when the reactants are mixed together, as when many acids are added to an alkaline substance (here the 'kinetic barrier' to be overcome is low). Other reactions take place at a significant rate only when the temperature is raised or a catalyst is present. Free energy is released when both types of reaction take place. That is to say they are both exergonic $(-\Delta G)$. The second type of reaction mentioned has to overcome a greater 'kinetic barrier' before it can take place. Supply of heat to the system allows it to do this, as will the introduction of a catalyst into the system. Thermodynamics tells us about the energy changes involved, but nothing about the rate of a particular reaction. For instance, two reactions may each be spontaneous and have the same overall free-energy change, but may reach equilibrium at different rates (see figure 4.1).

Molecules are able to react with each other only when they are sufficiently close together, so that in terms of molecular kinetics they may be said to collide. In the case of gases, molecules are considered merely to bump into each other; this is possible because the molecules are able to move randomly and independently of each other. The molecules of liquids, on the other hand, show diffusional movements of the solute molecules in the solvent.

We have already mentioned a type of reaction where there is a large 'kinetic barrier' to be overcome before the reaction will take place. Figure 4.2 shows a simple classical model of this situation. However, a more satisfactory way of looking at this is to consider that there is a threshold level of energy required for the reaction to take place. In cases of this type of reaction only a small proportion of collisions is effective. This assumes that the energy or speed of the average molecule is not sufficient for the reaction to take place. The molecules within a system have energy levels which show a particular distribution (see figure 4.3). At lower temperatures, relatively fewer molecules are at a higher energy level. However, if the temperature is increased, then the distribution of energy levels of the molecules changes as shown in figure 4.3. In situations where most mole-

33

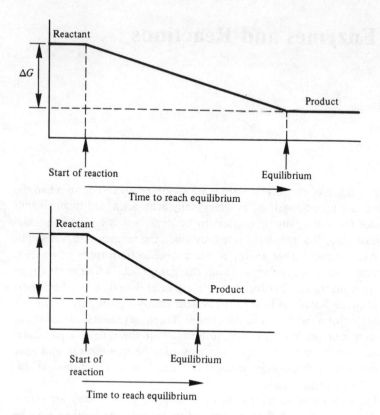

Figure 4.1   *Two spontaneous reactions with the same overall free-energy change but which reach equilibrium at different rates.*

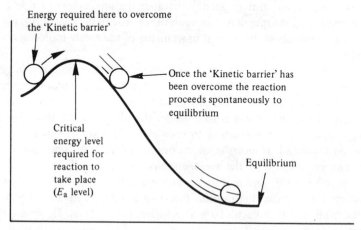

Figure 4.2

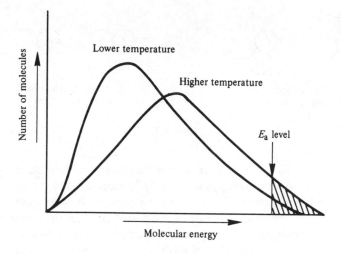

Figure 4.3

cules do not have sufficient energy to react, then they have not reached this critical energy level, which is referred to as the **activation energy** (E_a or $\Delta G^{\ddagger}$). In this way it can be considered that those molecules which are above the E_a level are **reactive**. That is to say, those molecules that are above the E_a level will be capable of reacting when they collide.

Now reconsider figure 4.2 which shows the simple model of an exergonic reaction. Although this reaction may have a highly negative $\Delta G^{\ominus}$ value, the reaction may occur very slowly. A useful analogy to make at this point may be to imagine a boulder near the top of a steep hill. The boulder has 'potential' energy which can be realised only by exerting extra energy on the boulder in order to take it over the top of the hill so that it can roll down.

It must also be remembered that the molecules of the reactant are held together with chemical bonds. These bonds must be broken before the component atoms can be rearranged to form the product molecules. Also, if you consider the positioning of the molecules when they collide, this must also be suitable even if the energy level is favourable as shown in figure 4.4. This means that there is an element of probability about the outcome of a collision, even when the energy levels are above the critical E_a level.

In living cells most chemical reactions are of the kind where there is a significant 'kinetic barrier'. This barrier formed by the activation energy is important, otherwise structural molecules of organisms would break down. As it is, biological systems have their own method of selective control of

35

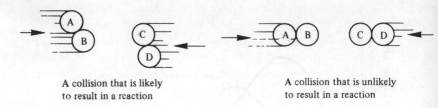

A collision that is likely
to result in a reaction

A collision that is unlikely
to result in a reaction

Figure 4.4

these processes. However, to cause one of these reactions to occur in living things, raising the temperature is not usually appropriate. Instead, enzymes are used and their function is to make the energetic situation suitable for the reaction to take place at a rate which is sufficient for the organism to remain alive. While most organisms have temperatures slightly above that of the environment, mammals and birds have body temperatures of about 35–44°C which are nearer to the optima of most enzymes and therefore reactions occur more rapidly.

Enzymes are biological catalysts and they behave in a similar way to inorganic catalysts, thus increasing the rate of a reaction. The current theory of the way that many catalysts work is that they offer a new pathway for the reaction which provides a lower free energy of activation (see figure 4.5). Consequently, without altering the energy distribution of the

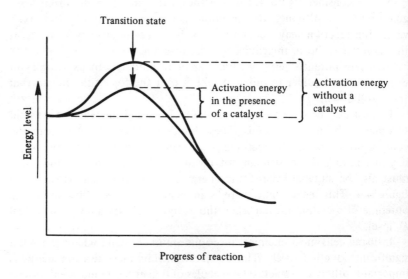

Figure 4.5

molecules, this has the effect of making more of the molecules in the system have an energy level which is now above the E_a level (see figure 4.6). Of course the reactions involving the catalyst and those not involving the catalyst each occur within the system at the same time, but the catalytic reaction involves so many more molecules at any one time that the rate of the reaction without the catalyst is insignificant.

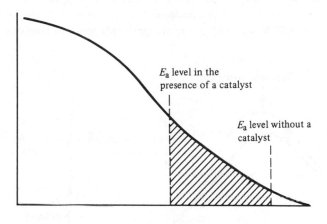

E_a level in the presence of a catalyst

E_a level without a catalyst

Figure 4.6

A simplified scheme for the mode of action of a catalyst of this type is

catalyst + reactant $\longrightarrow$ catalyst–reactant complex $\longrightarrow$ product + catalyst

The catalyst molecule is unchanged and may be used again and again in a similar reaction. Enzymes are catalysts which appear to function partly in this way, the substrate molecule becoming reversibly bound to the enzyme molecule before the product is formed. The rate of the reaction is usually affected by the concentration of the enzyme when the concentration of the substrate is not a limiting factor. In this way more enzyme–substrate complex is formed at any one time, thus speeding up the reaction. This is simply part of the current theory of enzyme action, but because the enzyme–substrate complex is both very unstable and very short-lived, proof of its identity is difficult to obtain.

Let us now return briefly to consider a simple model of a chemical reaction such as the conversion of A to B. This reaction will occur because a proportion of the molecules of A have sufficient energy at a particular instant to take on an activated condition. This activated condition is called the **transition state**. In the transition state there is a greater probability

37

that chemical bonds will be either made, or broken, to form substance B. The transition state occurs at the point on the 'kinetic barrier' when reactants are able to form products spontaneously. The rate of the reaction is consequently related to the proportion of molecules of A that are in the transition state.

If we stay with the reaction involving the conversion of substance A to B, then the free energy of activation of A $(\Delta G^{\ddagger})$ is the amount of energy required to convert all of the molecules in 1 mole of A to the transition state. Figure 4.7 shows some of the energetic relationships involved.

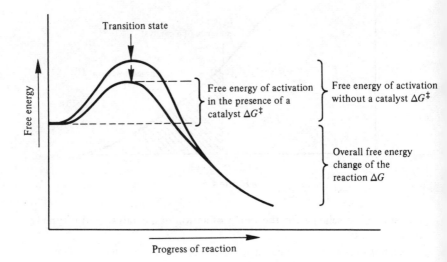

Transition state

Free energy

Free energy of activation in the presence of a catalyst $\Delta G^{\ddagger}$

Free energy of activation without a catalyst $\Delta G^{\ddagger}$

Overall free energy change of the reaction ΔG

Progress of reaction

Figure 4.7

Again it can be appreciated why the rate of a reaction will be increased by a rise in temperature. This is because a rise in temperature is a means by which the free energy of activation can be increased, since this increases the proportion of molecules that are able to take on the transition state.

In many reactions the rate is approximately doubled with each $10°C$ rise in temperature. This is called the Q_{10} **value**. This is sometimes written $Q_{10} \approx 2.0$. With enzyme-controlled reactions the Q_{10} value may vary with different enzymes.

It has to be realised that catalysts cannot make reactions occur that are energetically unfavourable since catalysts can only speed up the attainment of an equilibrium. Energetically unfavourable reactions do occur in biology, and these are able to take place only by being linked to energetically favourable reactions. An example is where redox reactions are used to provide the energy for ATP synthesis from ADP (see chapter 3).

Catalysts are said to be of two main kinds: 'homogeneous', as described above, where an intermediate is formed, and 'heterogeneous' where the mode of action is somewhat different. In this second type of action the catalyst is often referred to as a surface catalyst. Here the molecules of the reactant are held on to the molecule of the catalyst so that they are correctly aligned for maximum reactivity. This allows more effective collisions to take place. It appears that enzymes are capable of working both as homogeneous and heterogeneous catalysts. During the rest of this section it should become more clear how enzymes fulfil this function.

Enzymes have another important feature not shared by inorganic catalysts and this is that they are highly specific regarding the substrate that they will affect. This is important in biological systems because unless an enzyme affects only one particular reaction, so that each produces a specific product, then many reactions, each with different by-products, would occur and this would result in a wide range of unwanted or waste materials. Enzymes also have the ability to accelerate reactions to a quite incredible degree, far more so than inorganic catalysts. This may be as much as 10^6 or 10^7 molecules of substrate per minute for each enzyme molecule with highly active enzymes such as catalase (an enzyme which converts hydrogen peroxide to water and oxygen in most living tissues, or in the case of FMN or FAD, chapter 3). Enzymes consist either of proteins alone or proteins associated with other non-protein materials. These non-protein materials may be closely bound to the protein molecules, in which case they are referred to as **prosthetic groups**, or else, if the material is more loosely associated, then the term **co-factor** is often used (figure 4.8). Because enzymes are protein based they are affected by factors which are

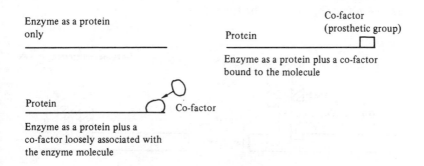

Enzyme as a protein only

Protein ——————— Co-factor (prosthetic group)

Enzyme as a protein plus a co-factor bound to the molecule

Protein ——————— Co-factor

Enzyme as a protein plus a co-factor loosely associated with the enzyme molecule

Figure 4.8

39

known to affect protein, such as denaturing at temperatures above 50°C, or by ions of heavy metals or they are affected by changes of pH. Most enzymes are active over a small range of pH and activity declines sharply outside these pH levels. Inside the cells pH is often controlled to remain at about the optimum for the enzymes.

Enzymes are often named by adding the suffix -ase after the name of the substrate or the reaction that they catalyse. For instance, lipases break down lipids or fats, and dehydrogenases are enzymes involved in reactions where hydrogen is removed (for example, those involving NAD^+ or FAD). Where co-factors are involved they may also be called **co-enzymes**. Without these co-factors then enzymes which depend on them become inactive. Cytochromes, mentioned in the previous chapter, are enzymes in which iron in the Fe^{2+} or Fe^{3+} state is present as the co-factor. Other dehydrogenase enzymes, also mentioned in the previous chapter, have co-enzymes such as NAD which is loosely associated with the enzyme and forms an enzyme–co-enzyme complex at the time of reaction, or FAD which is more tightly bound to the enzyme. It is interesting to note that many of the co-enzymes that have a central role in cellular energetics are vitamin B derivatives.

The catalytic activity of an enzyme is situated at a particular part of the molecule known as the **active site**. The highly specific nature of enzyme reactions suggests that enzyme action involves some sort of correspondence in the molecular structure between the enzyme and the reactant. In this way you can imagine a part of the molecule of the substrate molecule being temporarily attached to a corresponding and complementary part of the enzyme molecule (see figure 4.9). A simple and traditional analogy that has been used for some time is that of a lock and key. This type of analogy clearly shows how the correspondence between the two molecules relates to enzyme specificity, and the figure shows how an enzyme–substrate complex, similar to the catalyst–reactant complex mentioned earlier in this chapter, may be formed.

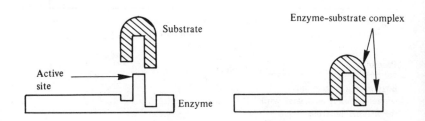

Figure 4.9 *Lock and key model of enzyme action.*

Although the 'lock and key model' seems to relate to a good deal of what is going on, a further modification is needed to explain the enzyme behaviour that has been observed. If you now consider that the complementary parts of the substrate and enzyme molecules do not correspond exactly but that they are at the same time both flexible, then in order to allow for the enzyme-substrate complex to be formed, changes in the shape of the active site of the enzyme have to occur to accommodate the substrate. The result of this would be to place extra 'stress' on the chemical bonds of both the substrate and enzyme. This would make the complex highly reactive. This theory is often referred to as the 'induced-fit hypothesis'. There is still much to find out about the nature of enzyme action and this is an area of considerable research activity.

Most of what we have been discussing here concerns models and analogies. These aids to thinking and understanding are frequently employed by scientists. However, it is essential that it is appreciated that these are only models and analogies and therefore they are unlikely to correspond to the real situation in all respects. As more evidence is obtained through careful investigation, so the models are modified and analogies changed. The area of enzyme biochemistry is a good example of scientific thinking of this kind, and even here we have discussed at least one case of a model being altered in an attempt to arrive at a better explanation of a current thory.

ENZYMES AND REACTIONS – SUMMARY

1. Some reactions take place at a significant rate at normal temperatures only when a catalyst is present. A 'kinetic barrier' has to be overcome for the reaction to occur.
2. Molecules in a system have energy levels which show a particular distribution, and only a certain proportion has an energy level (E_a) sufficient for them to react.
3. The activated condition where substrate molecules are more likely to react, that is, to make or break chemical bonds, is at the E_a energy level and is called the transition state.
4. Enzymes are used in biological reactions to make the energetic situation more suitable, thus speeding up the reaction.
5. Enzymes cause more molecules in the system to have an energy level above the E_a level.
6. Reaction rate increases with temperature because proportionally more molecules achieve the E_a level.
7. In many reactions the reaction rate approximately doubles with each $10°C$ rise in temperature. This is the Q_{10} value.

8. Catalysts and enzymes can only speed up reactions which are thermodynamically favourable in the first place.
9. Catalysts are of two kinds — (a) homogeneous, where an enzyme-substrate complex is formed which is transitory and (b) heterogeneous, where the substrate molecules are correctly aligned to increase reactivity.
10. Enzymes are highly specific regarding substrate or chemical bond.
11. Enzymes consist of either proteins or proteins with other materials.
12. Non-protein material loosely associated with the protein of an enzyme is called a co-factor or co-enzyme.
13. Most enzymes are active over a narrow pH range.
14. The part of the enzyme molecule where the catalytic activity is located is the active site.
15. A simple theory of enzyme action is the 'lock and key' hypothesis. This explains the high level of specificity of enzymes.
16. A modification of the 'lock and key' hypothesis is the 'induced-fit hypothesis' where complementary parts of the enzyme and substrate molecule do not correspond exactly. This means that for the parts to fit, they need to be flexibly moved to create an induced fit, thus causing stress on the chemical bond and making the complex more reactive.

5 Biological Membranes

Membranes are much more than the physical boundary of each cell. In eukaryotic (nucleated) cells, the different cellular organelles, such as the chloroplasts and mitochondria, are made of elaborate membrane structures. The membranes of different organelles and cells each have a characteristic structure and function. Wherever they occur, membranes are the site of a good deal of the activity of the cell and it is part of the purpose of this chapter to give an outline of some of the ideas concerned with the relationship between membrane structure and function. There are a number of functions associated particularly with membranes; these include the rather obvious need to keep materials in or out of the cells and at the same time to allow the passage of certain substances through selectively. Membranes in certain parts of cells are also important as the sites of energy-producing activities.

The study of the structure and functions of biological membranes is another good example of the way that scientific enquiry works. Over the years, different models of the structure of membranes have been put forward in an attempt to explain the very latest findings of investigators. As new observations have been made, so the structural model has been changed to account for this new information. Thus the structurel models have been, at each stage, only a current theory which has been adopted by workers until new information has required further modifications to be made to the model.

In recent years there has been considerable controversy over the nature and structure of biological membranes. Earlier models such as that proposed by Davson and Danielli in 1935, shown in figure 5.1a, contained elements that have been retained; these include the phospholipid bilayer and globular proteins. By the middle of the 1970s, one model, that proposed by Singer and Nicholson, had gained fairly general support as the most acceptable summary of current thinking. This is usually referred to as the **'fluid mosaic'** model of membrane structure and it may be helpful to begin by considering some of the important aspects of this model (see figure 5.1b).

43

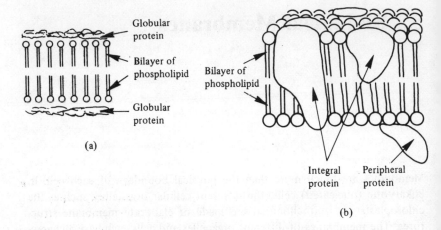

Figure 5.1 *(a) Davson and Danielli model of membrane structure. (b) Fluid mosaic model of membrane structure.*

All membranes appear to have certain features in common. For instance, they all contain lipids (most commonly phospholipids), proteins and carbohydrates; however, in different types of membranes these chemical constituents are present in different proportions. The carbohydrate molecules are either attached to lipids to form glycolipids or lipopolysaccharides, or with proteins to form glycoproteins.

The membrane lipids have two parts to the molecule. One part has chemical groups that are **hydrophobic** (water repellent) and the other has **hydrophilic** groups (with an attraction for water). In aqueous solutions this means that molecules of lipids naturally form double layers with the hydrophobic groups facing inwards (see figure 5.2).

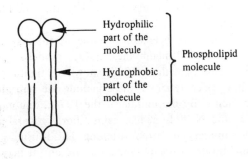

Figure 5.2 *Arrangement of phospholipids in a double layer in a membrane.*

44

The membrane proteins may penetrate and span the whole thickness of the membrane or they may be particularly associated with the upper or the lower surface of the membrane (as shown in figure 5.1b). However, it should be appreciated that the structure of the membrane appears to be dynamic; that is, it is not static and it seems that the proteins are constantly changing their positions in the membrane. The common analogy is that the proteins seem to float and move about in a sea of lipid.

The membranes of mitochondria and chloroplasts are also of particular interest; they are often referred to as **energy-transducing membranes** and they have a high proportion of protein in their structure. It is here that ATP synthesis takes place, brought about by enzyme complexes associated with the membrane (see chapter 3).

For living things to survive, the cells must be capable of taking energy and raw materials, including oxygen and food, from the environment. The waste materials, which become toxic if they accumulate, must be eliminated. The structure of membranes must allow for this two-way movement and one of the more obvious features is that materials pass in and out of the cell or organelle selectively (figure 5.3). The membrane is therefore a selective barrier between the inside and outside with the proteins forming the selective channels. The molecules and ions involved in these movements are so varied in size that simple diffusion methods are usually inadequate, therefore a range of mechanisms has developed in cells to take care of this.

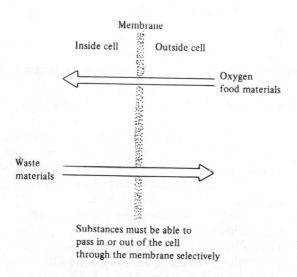

Figure 5.3

The movement of ions or molecules across membranes can occur in response to concentration gradients, electrical potentials set up on either side of the membranes (nerve cells), metabolic energy (kidney tubules) or a combination of these factors. The thermodynamics of systems involved in the movement of solutes across membranes is an important consideration. When a solution is diluted, the entropy of the system increases and there is a consequent decrease in free energy (see chapter 1). Conversely, as solutes are added to a solution, to make it more concentrated, the free energy increases. Free energy is therefore needed for a solute to move from a solution on one side of a membrane with a lower concentration of the solute into one on the other side, where the concentration is higher. For the same reasons there is a reduction of free energy when a solute moves into a solution of lower concentration. The energy implications here are that in the first case the process must be coupled to a system that releases free energy and in the latter it will occur spontaneously (see figure 5.4).

Ions are charged particles and there is often a difference in the concentration of these on either side of the membrane. This means that there is usually a difference in charge between the two sides of the membrane. As a result a **membrane potential** is produced ($\Delta \psi$). In some cells this is maintained at a high level and it has an important function, as in nerve or muscle cells (see chapters 9 and 10); on the other hand, in most cases it is minimised by selective exchanges of ions across the membrane; often the ions involved are potassium and sodium.

PASSAGE OF MATERIALS ACROSS BIOLOGICAL MEMBRANES

There are basically two ways in which materials move across membranes

(1) Passive, where the material is soluble in the hydrophobic part of the phospholipid membrane layer; this happens with uncharged molecules such as oxygen, water and carbon dioxide. Here, the mechanism may be by diffusion or **facilitated diffusion** using carriers which are not energy dependent.
(2) The other method involves energy-dependent carriers and it is commonly called **active transport**. It occurs in the protein part of the membrane (see figure 5.5). Transport proteins in cell membranes have many of the characteristics of enzymes although, strictly speaking, they may not be true enzymes. Certain proteins may be specific to one type of organelle, or even to certain types of organelles in one

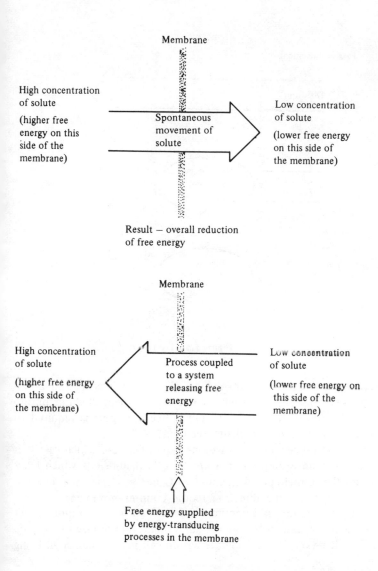

Membrane

High concentration
of solute

(higher free
energy on this
side of the
membrane)

Spontaneous
movement of
solute

Low concentration
of solute

(lower free energy
on this side of
the membrane)

Result — overall reduction
of free energy

Membrane

High concentration
of solute

(higher free energy
on this side of
the membrane)

Process coupled
to a system
releasing free
energy

Low concentration
of solute

(lower free energy on
this side of the
membrane)

Free energy supplied
by energy-transducing
processes in the membrane

Figure 5.4

kind of tissue. Particular membrane proteins, often known as transport proteins, such as the Na^+-K^+ exchange pump seem to interact specifically with particular materials and their purpose is to move molecules or ions across the membrane rather than to perform other functions more typical of enzymes.

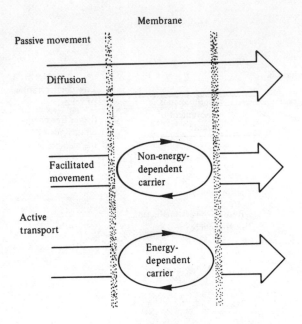

Figure 5.5

In summary, passive processes involving molecules or ions crossing a biological membrane usually occur when there is movement down a concentration or electrochemical gradient, and active transport is required to move materials against such gradients (see figure 5.6).

The transport process may involve one type of ion, in which case it is called a **uniport**; an example is the uptake of calcium ions which takes place across the mitochondrial membrane. Other processes may link the movement of ions in one direction to the counter-movement of a different type of ion; this is sometimes referred to as an **antiport** or an exchange, as in the case of Na^+ and K^+ movement in cells (see figure 5.7). These and other variations all operate within cell membranes in all living organisms.

(1) Passive Movement Through Biological Membranes

Some materials such as oxygen and water are able to pass through the cell membrane by simple diffusion through the phospholipid layer. The diffusion occurs as a result of concentration gradients between the two sides of the membrane. There is, therefore, a net movement of the material down its concentration gradient. This does not involve the cell directly in

48

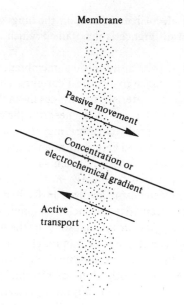

Figure 5.6

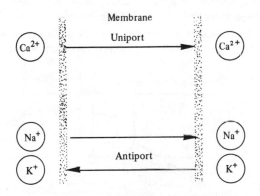

Figure 5.7

any expenditure of energy, although it should be appreciated that energy may have been expended in establishing and maintaining the gradient. The rate of movement of materials is dependent on the size of the concentration gradient on the two sides of the membrane. Thus, the greater the difference, the more rapid the diffusion rate. An example is the movement

49

of oxygen across the alveolar membrane in the lungs where by ventilation a large concentration difference is maintained which maximises diffusion rate.

Charged ions do not pass easily through membranes, as nearly all membranes show a high resistance to electrical currents. This has the advantage that it prevents leakage of materials, but it does mean that simple diffusion cannot be relied upon to take care of the needs of the cell.

Some forms of facilitated transport may be passive, as mentioned previously. The carriers involved here are not linked directly to any energy-releasing system. Substances such as certain simple sugars or amino acids are thought to be transported frequently into cells in this way. The explanations of the actual mechanisms involved in facilitated passive transport systems are somewhat speculative but include a number of theories, several of which could operate. One idea is that some of the proteins that traverse the membranes could create pores or minute channels through the membrane, thus allowing the selective movement of particular molecules and ions (figure 5.8). Recent investigations of membranes using high-power electron microscopy have provided supporting evidence for the existence of such pores in many types of membrane, showing how detailed structural investigations can help to support experimental biochemical findings.

Another suggested mechanism is that membrane transport proteins could pick up molecules or ions on one side of the membrane to form a complex, perhaps somewhat like an enzyme–substrate complex discussed in chapter 4. Then once the complex is formed, the combined molecule could be involved in movements, probably brought about by structural

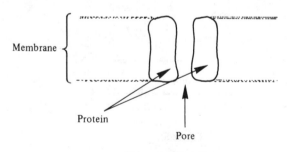

Membrane

Protein

Pore

(the pore is thought to
act also as a 'gate' which
opens only in response to
recognition of the approptriate
solute)

Figure 5.8

changes. These movements cause the molecule to 'flip' over across the membrane. The molecule or ion can then be released when it has passed across the membrane. As a result the carrier protein would then revert to its original state again and 'flip' back (figure 5.9).

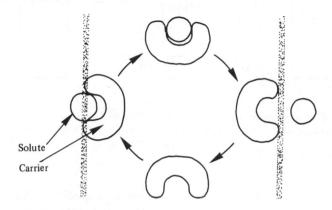

Figure 5.9 *Solute carried across the membrane by a carrier which attaches to the solute and then carries it to the other side.*

Yet a further theory is that in some cases membrane proteins that span the thickness of the membrane have molecules or ions that are to be transported bound to the protein on one side of the membrane. The action of binding of the molecule or ion causes a structural change (often referred to as a **conformational change**) in the protein. Such conformational changes can occur in protein molecules and involve movements of sub-units relative to each other. The effect of this change is to cause the molecule or ion to be moved to the other side of the membrane where it is then released (figure 5.10).

(2) Active Transport Mechanisms

It should be appreciated at the start that the mechanisms of ATP synthesis and ion transport are now seen to be closely linked, as will be shown later in this section.

All living cells expend energy in order to transport solute molecules or ions either in or out of the cell against a concentration gradient. By the processes of active transport, cells are able to maintain a very stable internal environment even when there are substantial changes in the external environment. Active transport is a process whereby ions or

51

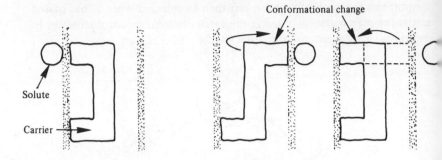

Figure 5.10 *Carrier molecule attaches to the solute and then, because of a conformational change, the carrier molecule moves so that the solute is transported to the other side of the membrane.*

molecules are accumulated against a concentration gradient. This type of transport results in an increase in the free energy within the system; it must in consequence be coupled to a process which makes free energy available. In other words, movement of solutes against a concentration gradient involves $+\Delta G$ values and these processes must be coupled to others that have even greater $-\Delta G$ values.

In order that charged ions or molecules should be able to move against a concentration gradient, they have to move against both the gradient of ions and of electrical charge. This electrochemical gradient requires even more energy to overcome it (see figure 5.11).

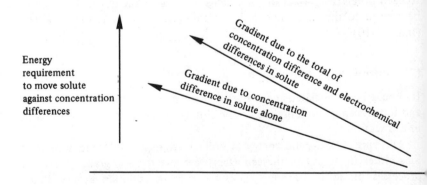

Figure 5.11

52

The membrane proteins involved in active transport, as mentioned before, strictly speaking may not be true enzymes, but they do have many properties in common with enzymes and two of them are of particular importance in the study of carrier mechanisms. These are the high level of specificity of the carriers and also the fact that their mechanism can be inhibited by specific agents. This latter property has been a useful tool in the study of the mechanisms involved. As an example of this, the use of specific inhibitors in investigation of the Na^+-K^+-ATPase pump will be discussed later. The proteins involved in active transport appear to have specific binding sites and they are also linked to energy-transfer systems. The direction of movement is always the same for a particular carrier.

There are two main types of active transport mechanism; one is referred to as **primary active transport** and the other as **secondary active transport**.

PRIMARY ACTIVE TRANSPORT

In this form of transport the energy is obtained from the hydrolysis of ATP by the electrons moving along an electron transport chain (see figure 5.12). One theory is that with an input of energy the carrier is able to form a complex with the molecule or ion that is to be transported. The complex formed between the carrier and the ion or molecule which is

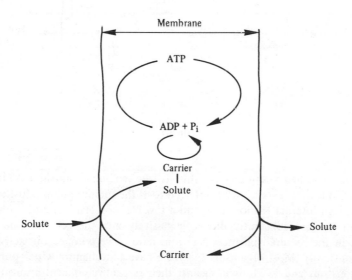

Figure 5.12

carried may then diffuse across the membrane; alternatively, the complex may rotate or change its structural state. The effect of this would be that the binding site will face the opposite side of the membrane, enabling the carrier to release the ion or molecule (see figures 5.13 and 5.14).

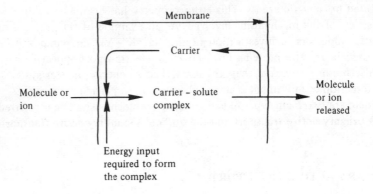

Figure 5.13

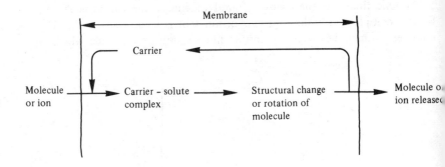

Figure 5.14

A well-studied example of this is the sodium–potassium–ATPase (N^+-K^+-ATPase) system, often referred to as the sodium pump. Most cells have a high internal K^+ ion level and a low Na^+ ion level, and as a result sodium ions are constantly diffusing passively into the cell. In order to maintain the required balance, Na^+ ions have to be pumped out actively and constantly. Most organisms therefore have a mechanism which pumps Na^+ ions out and K^+ ions in against their respective concentrations (see figure 5.15). This has special significance in the kidney tubule cells, erythro-

54

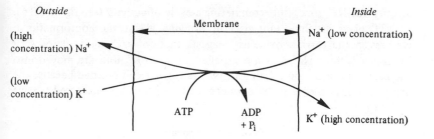

Figure 5.15

cytes and nerve axons (see chapter 9). Here as much as 70 per cent of the energy used in some cells, such as the kidney tubule cells, is used for active transport of this kind. This coupling of ATP hydrolysis to movement of ions against concentration gradients is an example of primary active transport.

Na$^+$-K$^+$-ATPase is an enzyme that hydrolyses ATP to ADP in the presence of Na$^+$ and K$^+$ ions. The enzyme is a large protein molecule that bridges the entire thickness of the membrane and it is responsible for the catalysis of Na$^+$ and K$^+$ ion movements and ATP hydrolysis. This enzyme is the basis of the sodium and potassium pump which is so important throughout cell function. In the process the two cations are each transported in one direction against their respective concentration gradients.

The current hypothesis is an interesting example of a biochemical theory which has gradually evolved over the years to explain a complicated phenomenon. It has, through a number of changes, attained its latest somewhat complex form but it does show how one current theory can link ion transport with ATP hydrolysis. The suggestion is that the Na$^+$-K$^+$-ATPase enzyme exists in two molecular conformational states referred to as E$_1$ and E$_2$. The E$_1$ form has a strong affinity for Na$^+$ ions and the E$_2$ form has an affinity for K$^+$ ions. Both forms can exist in phosphorylated states. When the form that has a high affinity for Na$^+$ ions is phosphorylated, Na$^+$ ions are transported across the membrane from the inside to the outside. There is then a conformational change in the enzyme and with this an accompanying change in affinity for Na$^+$ ions and K$^+$ ions, so that when K$^+$ affinity increases, Na$^+$ affinity decreases. Transport of Na$^+$ ions in one direction, and K$^+$ ions in the other, is brought about by conformational changes in the enzyme caused by phosphorylation and dephosphorylation coupled to ATP hydrolysis. Dephosphorylation of the enzyme causes K$^+$ transport from outside to inside. The whole cycle is completed by a change in the enzyme back to the structural form that has a high affinity for Na^+ ions, thus completing a cycle of activity (see figure 5.16). ATP plays an important part in bringing about the conformational changes in the carrier. Evidence suggests that for each molecule of ATP hydrolysed, three Na^+ ions are expelled and two K^+ ions are moved into the cells. The situation relating to Na^+ is further complicated because the membrane also seems to be permeable to Na^+ ions by facilitated passive

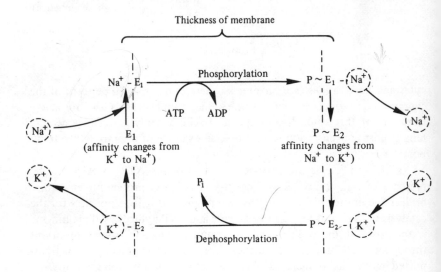

Figure 5.16

transport. In this way, Na^+ ions pass back into the cell down a concentration gradient using a carrier. One theory is that the carrier also has a binding site for another molecule such as glucose (see figure 5.17). In this way substances such as glucose may also be carried back into the cell with the Na^+ ions, even against their own concentration gradient. This type of system is referred to as a **co-transport** mechanism. This is an example of secondary active transport.

A technique that has been used to study membrane function has been to use specific inhibitors. In the case of Na^+-K^+-ATPase enzyme the glycoside ouabain has been used. This glycoside attaches to the enzyme on the outside of the membrane. The effect of this is that enzyme activity on the opposite surface of the membrane is inhibited and so also is ATP hydrolysis. This indicates that the transport of the two ions and ATP hydrolysis are linked.

56

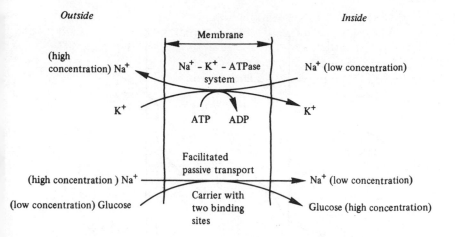

Figure 5.17

THE CHEMIOSMOTIC THEORY

An important theory which links active transport and membrane physiology with ATP synthesis is the **chemiosmotic theory** put forward by Mitchell in 1961. In recent years this theory has gained considerable support since it has a number of attractive features. In summary, the idea is that an electro-chemical gradient of hydrogen ions across a membrane is used as a source of energy for ATP synthesis.

Systems which produce proton gradients occur particularly in the inner membranes of mitochondria, the membranes of chloroplasts and the membranes of bacteria. The electrons originating from reduced co-enzymes (see chapter 3) or from the photolysis of water (see chapter 7) are passed along the chain of electron carriers associated with the membrane. Free energy is derived from the process and this is used to pump protons against the proton gradient to the opposite side of the membrane. In this way a proton gradient can be maintained against other forces (see figure 5.18).

The proton gradient thus created is used in certain ways. For instance, the backflow of protons across the membrane is used for ATP synthesis and, according to the chemiosmotic hypothesis, ATP synthesis and transport of materials across the membrane are closely related.

As a result of the proton gradient across the membrane created by the proton pump, this in turn may be coupled to a secondary active transport process. Here molecules are carried across the membrane coupled to

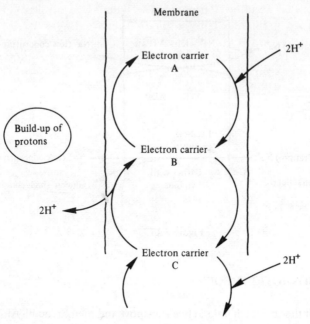

Outside   Inside

Figure 5.18 *Part of a sequence of electron carriers within a membrane, showing the pumping action of protons across the membrane to produce a proton gradient.*

protons moving back down their own concentration and electrical gradient (figure 5.19). A well-studied example has been the accumulation of lactose molecules in bacterial cells using this form of secondary active transport.

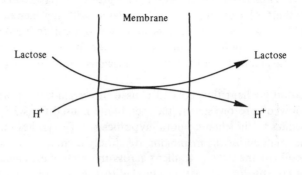

Figure 5.19

58

It is suggested that energy-transducing membranes have two types of protein complexes involved in energy conservation. One is called ATPase or perhaps it would be more helpful to refer to it as ATP synthetase. This enzyme is involved in the synthesis of ATP from ADP and inorganic phosphate (P_i); this functions with a positive input of free energy ($+\Delta G$). The second type of protein complex involved depends on which sort of energy-transducing membrane is under consideration. In mitochondria they form a chain catalysing the transfer of electrons ($-\Delta G$); similarly in chloroplasts they involve the use of energy made available from the absorption of light quanta.

Before the chemiosmotic theory had been put forward by Mitchell, no chemical intermediate could be found to link ATP synthesis with oxidation. With Mitchell's theory, the link between the two was provided by a proton gradient. The theory suggests that there are two types of proton pumps, one being driven by electron transfer or photon capture and the other by ATP synthesis (see figure 5.20).

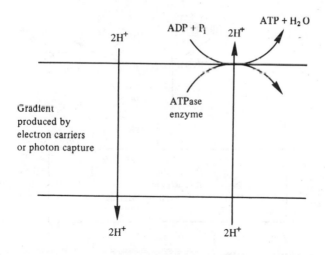

Figure 5.20

The electron and proton carriers have to be arranged within the membrane in a highly ordered way. As a result of this, protons are moved from one side of the membrane to the other when the electrons are passed through the carriers. There is a consequent build-up of protons to the outside and this produces a reduction of pH and negative charge on the inside of the membrane in mitochondria. The effect of this is to create a membrane electrical potential difference ($\Delta\psi$).

The existence of an electrochemical potential ($\Delta\psi$) for protons across the membrane is therefore an essential feature of the whole hypothesis; this potential can then be the means by which the free energy of electron transport is conserved. The build-up of protons on one side of the membrane is called the **proton motive force** (pmf or Δp). The proton motive force is due partly to the pH gradient across the membrane and partly to the electrical potential resulting from the activities of the electron-transfer chain in the membrane.

At first it was difficult to measure or even detect the pH differences or Δp, and this threw doubt on the whole theory. It has since been shown by careful investigation that the total electrochemical potential that occurs is of the appropriate order of magnitude to generate sufficient ATP.

It is an important feature of the membrane that it should also be impermeable to simple passive proton diffusion and that movement is by the means outlined in the chemiosmotic theory only.

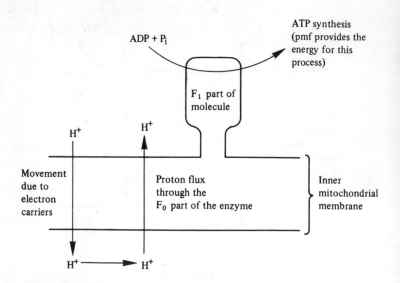

ADP + P_i

ATP synthesis
(pmf provides the
energy for this
process)

F_1 part of
molecule

H^+ H^+

Movement
due to
electron
carriers

Proton flux
through the
F_0 part of the enzyme

Inner
mitochondrial
membrane

H^+ ⟶ H^+

Build-up of
protons causes
a reduction of pH
which together with
an electrochemical gradient
results in the pmf

Figure 5.21

Another part of Mitchell's theory states that there are special channels in the membrane, through which protons pass. When this occurs the action of the ATP synthetase results in synthesis of ATP.

A special ATP synthetase occurring in two parts, and often referred to as F_1/F_0 ATP synthetase, is located in the inner mitochondrial membrane associated with special knob-like structures. The F_0 part of the ATP synthetase is within the membrane and forms a channel through which protons and electron movements take place, so providing the driving force for the phosphorylation of ADP by the F_1 part of the ATP synthetase complex within the knob (figure 5.21).

Classical experiments by Racker showed that if these knob-like projections, which are thought to contain the F_1 part of the complex, are removed then the membrane can still carry out the electron and proton movements but synthesis of ATP does not occur — thus providing evidence for the idea that proton and electron movement and ATP synthesis are carried out in separate parts of the complex. The whole theory is summarised in figure 5.21. It should be noted that the electron carriers of the respiratory chain (see chapter 3) function as a pump which transports H^+ across the inner membrane of the mitochondria.

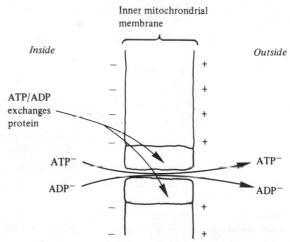

ATP molecules each have more negative charge than ADP molecules. The membrane potential has the inside negative in relation to the outside which means that outward transport of ATP and inward transport of ADP occur. This results in the overall movement of one negative charge from the inside to the outside with each ATP/ADP molecule exchanged.

Figure 5.22

61

ATP is formed on the inside of the mitochondria but it is required outside within the cell. This requires an ATP/ADP exchanger which appears to be a protein in the membrane. The basis for the exchange process seems to be that ATP molecules are more negatively charged than ADP and that the existing membrane potential favours ATP movement to the outside of the membrane (see figure 5.22). The actual mechanism may be by carrier or a pore mechanism but this will not be considered further here.

The relationship between the energetics of cells and the functioning of the membranes is closely interconnected, so that to study one necessarily involves consideration of the other. Developments in understanding of membrane physiology have automatically led to increased knowledge of energy conversion in biology and vice versa.

BIOLOGICAL MEMBRANES – SUMMARY

1. Membranes are more than the physical boundaries of cells. They are the site of much cellular activity. They are concerned with energy production in certain parts of cells and also with selection of materials entering or leaving cells or organelles.
2. Membranes contain lipid compounds, proteins and carbohydrates in particular combinations and patterns.
3. The fluid mosaic model is a widely held view of membrane structure, with a structure which is dynamic.
4. Material can move either by passive or active means across a membrane.
5. Passive means involve diffusion or facilitated diffusion using non-energy-dependent carriers. This occurs as a result of concentration or electrochemical gradients.
6. Active means use energy-dependent carriers and these move materials against concentration gradients. The protein parts of the membrane are involved in active transport.
7. If only one ion is involved in a transport process it is called a uniport, for example, calcium ions. Where there is an exchange, as in Na^+ and K^+, it is an antiport.
8. There are several explanations for facilitated transport. One is that proteins form pores; another is that carrier proteins are involved in structural or conformational changes which move the ion or molecule across the membrane.

9. For charged ions to move across a membrane against a concentration gradient they must move against the gradient of ions and electrical charge.

10. Membrane proteins have a good deal in common with enzymes. They are very specific and they can be inhibited by specific agents.

11. A theory of active transport is that the carrier forms a complex with the material to be transported. A familiar example is the Na^+-K^+-ATPase system which is coupled to ATP hydrolysis.

12. The current hypothesis is that Na^+-K^+-ATPase exists in two states, E_1 and E_2. E_1 has an affinity for Na^+ and E_2 for K^+. When E_1 is phosphorylated it transports Na^+ across the membrane. When Na^+ are released, a change occurs to E_2 and it attaches to K^+. Dephosphorylation causes K^+ to be transported back. When K^+ are released, E_2 changes to E_1.

13. ATP is involved in the changes from E_1 to E_2. For each molecule of ATP hydrolysed, three ions of Na^+ are expelled and two ions of K^+ taken in.

14. Specific inhibitors have been used to study membrane transport. Ouabain has been used for Na^+-K^+-ATPase and with it the linked transport of Na^+ and K^+ as well as ATP hydrolysis have been demonstrated.

15. The chemiosmotic theory links active transport with ATP synthesis. Here, an electrochemical gradient of hydrogen ions is created by the redox reactions that occur in the membrane. H^+ are moved to the outside and there is a membrane potential produced. This potential energy is used for ATP synthesis.

16. It is suggested that energy-transducing membranes have two sorts of protein complex: ATP-synthesising enzyme and the electron carriers.

17. Mitchell's theory suggests that two types of proton pump are involved, one with electron transfer and the other with ATP synthesis.

18. A special ATPase called F_1/F_0 occurs in the inner mitochondrial membrane. The protons pass through the F_0 part and phosphorylation of ADP to ATP occurs in the F_1 part.

19. An ATP/ADP exchanger is needed to move ATP out of, and ADP into, the mitochondria.

20. Secondary active transport involving movement of molecules like lactose coupled to movement of protons down their concentration gradient has been studied in bacterial cells.

6 Metabolic Pathways of Respiration

The next two chapters are concerned with two metabolic pathways which occupy a position central to cellular energetics. These are respiration and photosynthesis. Some detailed discussion of the actual pathways is given here, not so that they may be memorised, but rather that some idea may be given of the way that these complex biochemical manoeuvres fit in with basic ideas of energy and thermodynamics.

SOME METHODS OF INVESTIGATION

Before embarking on a discussion of the actual metabolic pathways of respiration in living cells, it may be helpful to digress briefly to consider some of the methods of investigation that have been developed by biochemists working in this field. The discoveries that have been made in this area are remarkable, and to a large extent they are the result of the ingenuity of workers in overcoming some of the difficult technical problems encountered.

One of the first problems faced by researchers in cell respiratory metabolism is that within cells a large number of reactions are taking place and only some of these are directly part of the respiratory process. The initial problem then is to disentangle the relevant reactions. Also, when tissues are separated from the body of an organism, they are at the same time separated from the blood supply carrying oxygen to the cells. Thus, in order to overcome this difficulty, thin slices of tissue may be used for study and in this way the oxygen requirements of the cells may be satisfied by diffusion. Although this technique has the advantage that it causes disruption to the minimum number of cells, it is perhaps less useful than cell preparations where mincing, grinding or fine homogenisation are used to break open the cells. Here, it is the intention to release the contents of the cells, including such organelles as the mitochondria. After this, using other methods, it is possible to separate the various cell components so that they can be studied in isolation. It was in 1948 that Hogeboom,

Schneider and Palade first used homogenisation coupled with separation techniques involving differential centrifugation to prepare extracts of mitochondria.

Two main methods of separating organelles and other cell fractions have been developed. One is **density gradient centrifugation** and the other is **differential centrifugation**. Both methods are based on the principle that different organelles and other cell fractions have a different density, and as a consequence, centrifugation can be used to separate one from the other.

In the case of density gradient centrifugation (see figure 6.1), a medium is prepared in a centrifuge tube so that it forms a gradient, with the most dilute medium at the top and the most concentrated at the bottom. The cell homogenate is layered carefully on the surface. When it is centrifuged,

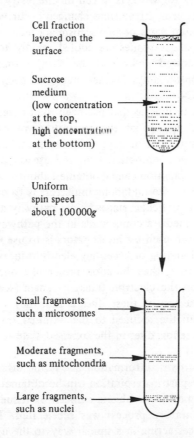

Cell fraction
layered on the
surface

Sucrose
medium
(low concentration
at the top,
high concentration
at the bottom)

Uniform
spin speed
about 100000*g*

Small fragments
such a microsomes

Moderate fragments,
such as mitochondria

Large fragments,
such as nuclei

Figure 6.1 *Density gradient centrifugation.*

the density gradient that has been set up does not change, but the fractions of the cell homogenate move down within the medium and remain at a level where their own density is the same as the surrounding medium. In this way, separation of organelles occurs, and the depth to which they travel for a particular centrifugation spin speed can be precisely determined. Particular parts of the cell fraction can then be withdrawn using a hypodermic syringe.

When differential centrifugation is used, the homogenate is placed in a uniform medium with no density gradient and consecutive centrifugations with increasingly greater spin speeds are used (see figure 6.2). With each spin, the sediment produced contains progressively less-dense fragments so that at known speeds with media of a particular density, the sediment will be made up of a particular part of the cell fraction.

A further difficulty for workers in cell biochemistry is that soon after the death of an organism, irreversible changes occur within the tissues. However, if the required parts of an organism are removed quickly after it has been killed, and these tissues are cooled rapidly to nearly $0°C$, the enzyme activity is slowed down and the changes that occur following death are reduced considerably. This technique has been widely employed in experimental work.

Respirometry is a technique that has been used extensively to investigate the biochemical nature of the respiratory process. This method usually involves measurement of oxygen uptake from living materials, and if this is carried out in conjunction with the use of selective metabolic enzyme inhibitors, information can be obtained about the initial substrates used. Also, the effect of the metabolic inhibitors is to cause the accumulation of materials at particular stages in the pathway and this can assist identification of intermediate compounds in the pathway (see figure 6.3).

Yet another approach used by investigators is to use **redox dyes**. These dyes are capable of donating or accepting electrons in oxidation or reduction reactions. Therefore, they function rather like one of the naturally occurring chemicals in the electron transport chain (see chapter 3). One great advantage in the use of these dyes is that they also change colour when they move from the reduced to the oxidised state or back again. Methylene blue is a redox dye; in the oxidised state it is blue and when reduced it is colourless.

Using these redox dyes, information about the effects of substances which inhibit or take part in respiration can be obtained. For instance, it was first found that methylene blue is reduced by tissue preparations and in this way the respiratory process was accelerated. This suggested to workers that substances acting in a similar way to the dye may be naturally present in respiring cells. A further useful discovery was that the

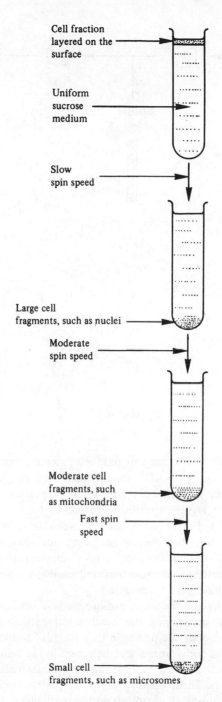

Figure 6.2 *Differential centrifugation.*

67

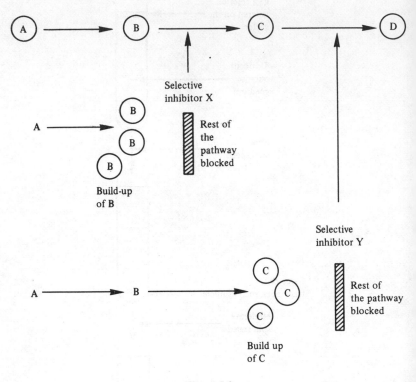

Figure 6.3

difference between the reduced or oxidised form of some naturally occurring substances such as NAD^+ can be identified by the characteristic changes that occur in their light absorption properties. Such substances often absorb light at very specific wavelengths; these are referred to as absorption bands and these bands differ between the reduced and oxidised forms. The use of this phenomenon has been made possible particularly by **spectroscopy** and later by the more refined spectrophotometer. Through this technique, patterns of reduction and oxidation, particularly of co-enzyme systems, have been investigated.

Chromatography and the use of **radioactive isotopes** have also played an important role in unravelling the biochemical pathways involved. With isotopes, workers adopt the principle that atoms of isotopes behave in the same way as the normal atoms and take part in the same chemical reactions. However, at the same time these radioactive materials can be detected in minute amounts. Thus, when isotopes of a particular element are introduced into an organism, all chemicals within particular pathways may have the isotope incorporated within them. They will then be said to be 'labelled'. With subsequent analysis based on separation using chromatography, useful information can be obtained. If the analysis is carried out at different times after introduction of the isotope into the system, different substances may be labelled, thus providing evidence for a particular sequence. Sometimes the difference in time from one step to another in the pathway may be very short indeed (see figure 6.4).

X - C*   Shown when analysed after a very short time

X - C* ⟶ Y - C*   Shown when analysed after a slightly longer time

X - C* ⟶ Y - C* ⟶ Z - C*   Shown after a longer time still

(C* radioactively labelled element)

Figure 6.4

RESPIRATION

The pathways to be discussed in this section are based on painstaking investigation over many years by a large number of workers using the techniques mentioned previously. The processes have been the subject of investigations which have achieved classic status and are discussed later in this chapter. To start with, we will consider some of the basic information involved. Broadly, respiration can be said to involve the chemical breakdown of fuel molecules and the recovery and conservation of part of the chemical energy as ATP.

A convenient place to begin discussion of the cellular respiratory pathway is with glucose. The whole biochemical pathway can be divided into two main parts. The first is **glycolysis** and this involves the anaerobic breakdown of glucose to pyruvate (more recently referred to as 2-oxopropanoic acid). Glycolysis may then either be linked to **aerobic catabolism** of the pyruvate using molecular oxygen or it may lead on to one of a number of different **anaerobic pathways** (see figure 6.5).

Aerobes obtain most of their energy from that part of the process which involves oxidation of organic compounds using molecular oxygen;

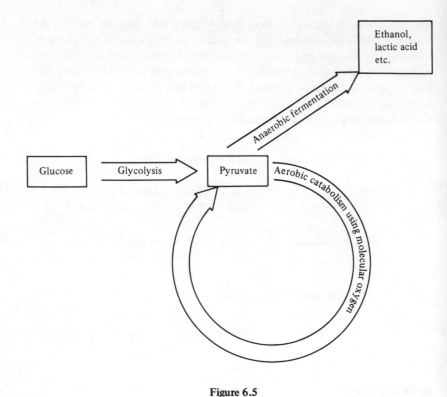

Figure 6.5

the energy here comes from redox reactions as outlined in chapter 3. Anaerobes utilise a different form of redox reaction.

The energy production of the different parts of the respiratory process are summarised in figure 6.6.

GLYCOLYSIS

This occurs in the **cytosol** (liquid part of the cytoplasmic solution) of cells. It does not require membrane structures and it appears to occur in all living things.

The overall reaction of glycolysis is exergonic and can be written thus

$$\text{glucose} \longrightarrow 2 \text{ pyruvate molecules}$$
$$(\Delta G^{\ominus} = -135 \text{ kJ mol}^{-1})$$

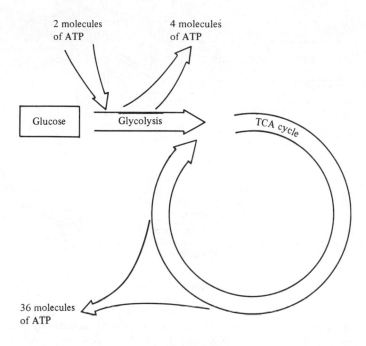

2 molecules
of ATP

4 molecules
of ATP

Glucose | Glycolysis | TCA cycle

36 molecules
of ATP

Figure 6.6

At the same time ATP is formed from ADP and phosphate during glyco-lysis and this is an endergonic process.

$$ADP + P_i (PO_4^{3-}) \rightarrow ATP$$
$$(\Delta G^{\ominus} = + 61 \text{ kJ mol}^{-1})$$

It can be seen here that there is quite sufficient free-energy change from glycolysis to drive ATP synthesis. However, glycolysis takes place in a series of steps, each catalysed by different enzymes, and each step involves a relatively small change in standard free energy.

The pathway begins with glucose and following on from this all of the intermediate compounds have a phosphate group attached to them and they are therefore referred to as **phosphorylated** organic compounds. The phosphate groups of each compound form the binding site for the enzyme-substrate complexes which are necessary for the reactions to take place. Also, the phosphate groups have an important part to play in the process of energy conservation and ultimately they become the terminal phosphate groups of ATP. The pathway occurs in three main stages (see figure 6.7). In outline, glycolysis can be divided into a number of stages. In the first

71

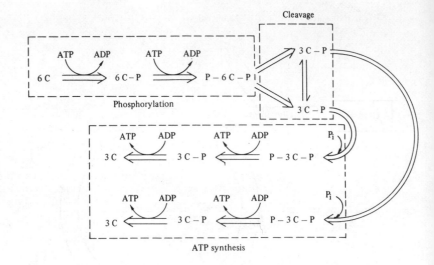

Figure 6.7

stage the glucose, a six-carbon compound, is phosphorylated by adding first one and then another phosphate group. To achieve this, ATP itself is used. Following on from this, in the second stage, the six-carbon phosphorylated compound is split to form two phosphorylated carbon compounds each with three carbon atoms. These three-carbon compounds are further phosphorylated and then they are used to generate two ATP molecules.

PHOSPHORYLATION

As previously mentioned, the first steps of glycolysis involve converting the six-carbon (hexose) glucose into a doubly phosphorylated state. To achieve this, glucose is first converted to glucose-6-phosphate using one molecule of ATP. After this, there is some reorganisation of the molecule to form fructose-6-phosphate which is then further phosphorylated to fructose-1:6-diphosphate. Fructose is an isomer of glucose — in other words it contains the same elements in the same proportions but in a different structure; so the interconversion is a fairly simple matter and it involves a relatively small free-energy change ($\Delta G^{\ominus} = 1.7$ kJ mol^{-1}).

The reason why inorganic phosphate cannot be used instead of ATP is that thermodynamically this would be unfavourable. In the reaction where glucose is phosphorylated using ATP, the free-energy change is

−16.7 kJ mol^{-1} (see table 2.1, chapter 2) and this reaction appears to be thermodynamically very favourable; so much so, that most of the glucose that enters the cell undergoes this reaction. This has the added advantage that in the phosphorylated form, the sugar molecules are unable to pass back across the membrane out of the cell.

The conversion of the phosphorylated glucose to fructose has the effect of creating a more suitable arrangement for the attachment of the second phosphate group. The first phosphate is attached to the sixth carbon atom of the glucose molecule by a hydroxyl group (–OH). However, in glucose, the first carbon atom to which the second phosphate is going to be attached is not part of a hydroxyl group, and therefore it is much less suitable. In fructose the corresponding carbon atom is part of a hydroxyl group (see figure 6.8).

SPLITTING THE SIX-CARBON DIPHOSPHATE

Following on from the initial stages described so far, the fructose-1:6-diphosphate is then split by a specific enzyme (aldolase) to form two sugars, each with three carbon atoms. These are also isomers of each other and readily interconvertible, the isomerisation reaction being brought about by the enzyme triose phosphate isomerase (see figure 6.9). Here although the equilibrium of this reaction is towards dihydroxyacetone phosphate formation ($\Delta G^{\ominus} = 7.5$ kJ mol^{-1}) the reaction actually moves in the opposite direction because of continual removal of glyceraldehyde-3-phosphate by subsequent reactions in the pathway. As has been pointed out in chapter 1, the relative concentrations of substrate and product are very significant and can strongly affect the direction in which a reaction takes place. At this stage it should be remembered that in the whole process so far, two molecules of ATP have been used for each molecule of glucose fed into the system.

ENERGY-YIELDING REACTIONS

The final part of the glycolytic sequence includes the energy-yielding reactions. However, it is an important principle that the electron transport chain mentioned in chapter 3 is not involved. In the first step, ATP synthesis is linked to an oxidative reaction; this occurs in two stages, one being the reduction of NAD^{+} and this is followed by phosphorylation of ADP. The oxidative reaction involving glyceraldehyde-3-phosphate is

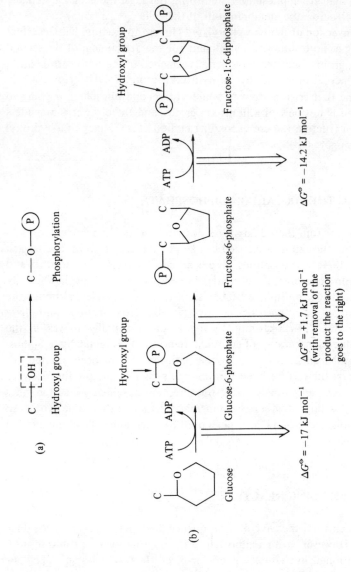

(a)

$$C \longrightarrow \boxed{OH} \qquad \xrightarrow{\hspace{2cm}} \qquad C \longrightarrow O \longrightarrow \text{P}$$

Hydroxyl group Phosphorylation

(b)

Glucose

Glucose-6-phosphate

Fructose-6-phosphate

Fructose-1:6-diphosphate

Hydroxyl group

Hydroxyl group

$\Delta G^{\ominus} = -17$ kJ mol^{-1}

$\Delta G^{\ominus} = +1.7$ kJ mol^{-1}
(with removal of the product the reaction goes to the right)

$\Delta G^{\ominus} = -14.2$ kJ mol^{-1}

Figure 6.8

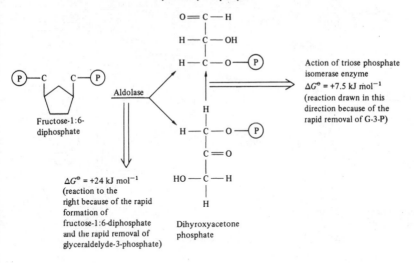

Glyceraldehyde-3-phosphate

$$O = C - H$$
$$H - C - OH$$
$$H - C - O - \text{(P)}$$

Fructose-1:6-diphosphate

Aldolase

Action of triose phosphate isomerase enzyme
$\Delta G^\ominus = +7.5$ kJ mol^{-1}
(reaction drawn in this direction because of the rapid removal of G-3-P)

$$H - C - O - \text{(P)}$$
$$C = O$$
$$HO - C - H$$
$$H$$

$\Delta G^\ominus = +24$ kJ mol^{-1}
(reaction to the right because of the rapid formation of fructose-1:6-diphosphate and the rapid removal of glyceraldelyde-3-phosphate)

Dihyroxyacetone phosphate

Figure 6.9

highly exergonic and this is enough for both NAD$^+$ reduction and ADP phosphorylation.

A significant aspect of this reaction related to the conservation of energy is the way that the oxidation is coupled to formation of a phospho-anhydride bond; this is an important feature of high-energy compounds (see chapter 2). This occurs with the first carbon atom of the 1,3-diphos-phoglycerate molecules and this subsequently allows the formation of ATP by transfer of the phosphate to an ADP molecule (see figure 6.10). Prior to this, a high-energy thioester bond was formed between the enzyme and attached substrate. Thioester bonds are important in respiratory processes since they have a high free energy of hydrolysis (see later in this chapter on the TCA cycle). Glycolysis is essentially anaerobic, therefore the oxidation reaction is achieved not by the addition of oxygen, but by the addition of water and the subsequent removal of hydrogen, thus leaving the oxygen behind. The removal of the protons and electrons is carried out by NAD$^+$, which forms a complex with the enzyme when in the oxidised form and frees itself when in the reduced state.

The effect of these chemical manoeuvres is to couple the exergonic reaction of the oxidation of glyceraldehyde-3-phosphate ($\Delta G^\ominus = -43$ kJ mol^{-1}) to the even more endergonic reaction which is the formation of the phosphoanhydride bond ($\Delta G^\ominus = +49.3$ kJ mol^{-1}). However, the overall result is that the coupled reactions occur in the direction of the

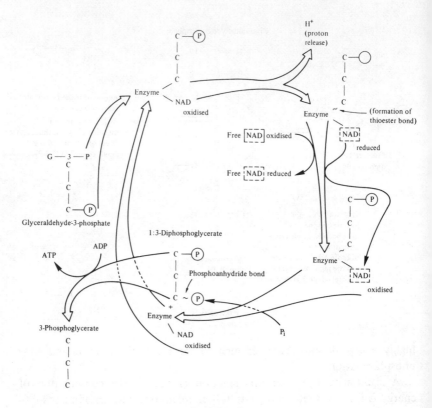

Figure 6.10

next step in the sequence. In this step in the pathway, the phospho-anhydride bond between the first carbon atom of the 1,3-diphospho-glycerate is transferred to ADP to produce ATP ($\Delta G^{\ominus} = +30.5$ kJ mol^{-1}) and 3-phosphoglycerate.

The final phase in the pathway involves the formation of another molecule of ATP. Here the phosphate group which is linked to the third carbon atom of the 3-phosphoglycerate is involved. The free-energy change that occurs in the conversion of this bond to the phosphoenol bond of the phosphoenol pyruvate that is formed, is achieved in several steps ($\Delta G^{\ominus} = 13.8$ kJ mol^{-1} to 30.5 kJ mol^{-1}). This can be looked upon as the conversion of a low-energy bond to a relatively high-energy phosphoenol (see figure 6.11).

In the first of the steps, the phosphate of the 3-phosphoglycerate is moved from the third to the second carbon of the molecule to form the compound 2-phosphoglycerate. This is a slightly endergonic reaction

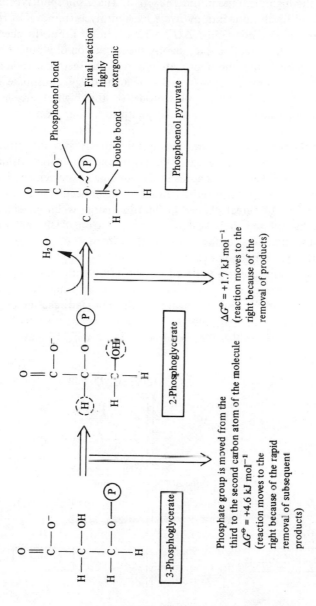

Figure 6.11

$(\Delta G^{\ominus} = 4.6 \text{ kJ mol}^{-1})$; however, the reaction moves in this direction because of the highly exergonic final reaction. The 2-phosphoglycerate is then converted to phosphoenol pyruvate by hydrolysis (removal of H_2O). This again is slightly endergonic $(\Delta G^{\ominus} = 1.7 \text{ kJ mol}^{-1})$, but the chemical advantage is that this reaction does involve the formation of a double bond and this means that now the phosphate group on the second carbon atom has one of the important characteristics of high-energy phosphate compounds. Phosphoenol pyruvate shows one of the highest free-energy changes known, when hydrolysis of the phosphate group occurs $(\Delta G^{\ominus} = -61.9 \text{ kJ mol}^{-1})$.

The penultimate step in the glycolytic pathway has the effect of bringing about a significant change in the distribution of energy within the molecule; this is achieved by changes in the bonding of oxygen and hydrogen with the second and third carbon atoms of the molecule.

As a result of the great change in the free energy when phosphoenol pyruvate is hydrolysed and ATP is synthesised from ADP and organic phosphate, the reaction readily moves in the direction of ATP synthesis. Consequently, phosphoenol pyruvate is converted to pyruvate with the production of a molecule of ATP (see figure 6.12). Synthesis of ATP from ADP has a free-energy change of about $+31 \text{ kJ mol}^{-1}$ and therefore the overall difference in free-energy change for this coupled reaction of phosphoenol pyruvate hydrolysis and ATP synthesis is $-61.9 \text{ kJ mol}^{-1}$ and $+31 \text{ kJ mol}^{-1}$ respectively, that is, about -30 kJ mol^{-1}, quite enough

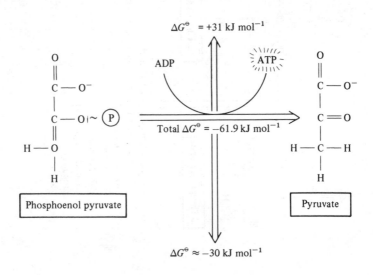

Figure 6.12

to drive the entire reaction. This type of direct linking of ATP synthesis, with a reaction in a metabolic pathway rather than having electron transport chains, is referred to as **substrate-level phosphorylation**. The entire pathway may be now seen as shown in figure 6.13. If we then look back to the oxidation of 3-phosphoglyceraldehyde, we see that it is achieved by the removal of electrons which are passed to NAD^+. This reduced electron carrier is re-oxidised in anaerobic systems by using it to reduce the pyruvate formed at the end of the glycolytic pathway and in this way, lactate is formed (see figure 6.14) or some other final breakdown product. This reaction is thermodynamically favourable ($\Delta G^{\ominus} = -25$ kJ mol^{-1}).

Lactate is the final product of a number of anaerobic systems including cheese-making and yoghurt-making by bacteria, as well as in the skeletal muscles of animals. However, another common pathway in anaerobic systems, which involves the re-oxidation of NADH, is alcoholic fermentation; this is a fundamental reaction of baking and brewing (see figure 6.15).

Glycolysis and fermentation release relatively little energy compared to aerobic processes. For instance, within the lactate molecule the free energy that remains is 2671 kJ mol^{-1} and only a small amount of the free energy of the original glucose molecule has been conserved within the ATP molecules formed. Thus, in order to conserve the same amount of energy by anaerobic means as by aerobic means requires a good deal more of the initial fuel material.

One reason why anaerobic systems produce much less energy per molecule of glucose than aerobic systems is that, generally, only a limited amount of oxidation of organic compounds is able to take place without oxygen being involved. In the anaerobic process on the other hand, each step involving oxidation of an organic compound, where electrons are removed, must be linked to reduction, where electrons are added to another organic compound which is usually closely related to the previous compound. The change in free energy which occurs as a result of this type of linked oxidative–reductive reaction is small and relatively little energy can be conserved in ATP. Most of the free energy is still within the final product of the pathway at the end of the process.

ALTERNATIVE PRIMARY SUBSTRATES

Instead of glucose, a number of other substrates are often used as initial sources of fuel within cells. These include fructose, glycogen and starch, and although it is not the intention to go into detail here, it should be understood that these alternative compounds usually undergo a series of

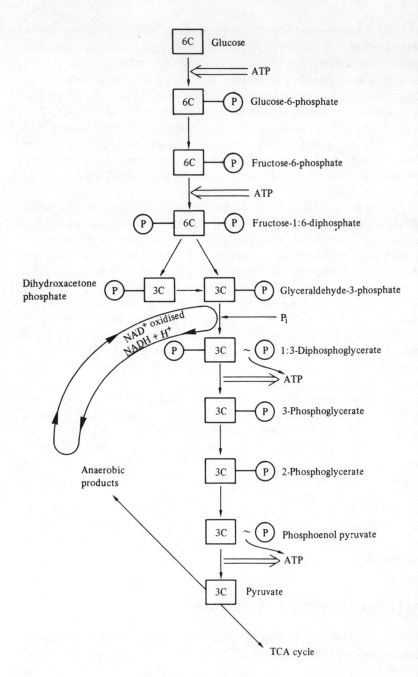

Figure 6.13

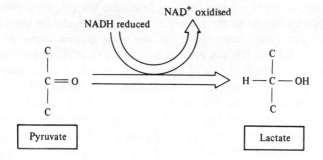

Figure 6.14

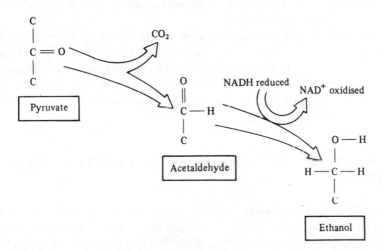

Figure 6.15

reactions which convert them into compounds which can then be introduced into the glycolytic pathway described previously.

For instance, fructose and ATP form fructose-6-phosphate and ADP. This fructose-6-phosphate then goes on in the normal way in the glycolytic pathway (see figure 6.16).

Figure 6.16

Very commonly the initial starting point for the glycolytic process is either glycogen or starch; these are the most common storage compounds of animals and plants respectively. In this way, glucose forms an intermediate substrate and it is not present to a large extent as a free compound. Carbohydrate is usually stored as a long-chain polysaccharide, rather than glucose, since this has the effect of reducing the osmotic effects of the storage compounds on the cells. Also the polymer (long repeated chain) is capable of bringing about a greater free-energy change when it is broken down than the equivalent amount of glucose. By this means, the yield is an extra ATP molecule for each glucose unit (see figure 6.17). The glycogen is broken down to release glucose units to which inorganic phosphorus is combined and this forms glucose-1-phosphate and then glucose-6-phosphate which enters the glycolytic pathway without the input of ATP for the first phosphorylation reaction.

It may be helpful to point out at this stage that in texts describing the respiratory process, some refer to organic acids such as pyruvic acid and lactic acid, while others refer to anions like pyruvate and lactate. It seems that at the pH levels that exist in animal and plant cells, these organic acids are in the ionic form and therefore there are good reasons for adopting the naming system used here.

AEROBIC PRODUCTION OF ATP (THE TRICARBOXYLIC ACID CYCLE)

In aerobic systems, certain carbon atoms of the organic compounds involved are oxidised to carbon dioxide; electrons are also removed and transferred to oxygen via a chain of carriers (see chapter 3). The oxygen acts as the final acceptor of electrons and in this way it causes continuous re-oxidation of the reduced co-enzyme molecules in the chain. In the aerobic system the pyruvate is not reduced to lactate or ethanol, but instead the carbon of the pyruvate is subjected to oxidation in a series of steps known as the **tricarboxylic acid cycle** (TCA cycle).

The TCA cycle can be considered to begin with the compound acetyl-co-enzyme A (acetyl-CoA), which consists of the acetate of the two-carbon compound formed by the oxidative decarboxylation (decarboxylation is loss of carbon dioxide from the carboxyl group) of pyruvate, which is itself part of a more complex nucleotide-based compound which makes up co-enzyme A (see figure 6.18).

The acetate (2-C) of the acetyl-co-enzyme A is then transferred to a four-carbon compound, oxaloacetate, which forms a six-carbon compound, citrate. Following on from this, in a cyclic series of reactions, the citrate

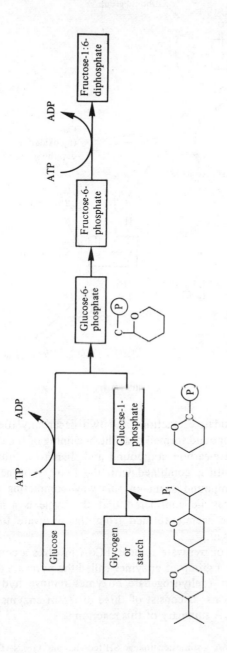

Figure 6.17

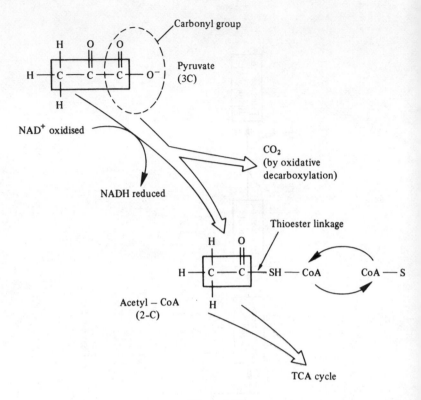

Figure 6.18

undergoes four oxidative reactions and two decarboxylation ones. Thus the six-carbon compound formed near the beginning of the cycle is broken down first to a five-carbon compound and then to a four-carbon compound which in turn is combined with the two-carbon acetate to form the six-carbon compound again—in this way completing the cycle (see figure 6.19). It may be considered that the cycle is a mechanism for breaking down the acetate formed from the pyruvate to yield carbon dioxide and hydrogen.

The conversion of pyruvate to acetyl-CoA involves a complex series of reactions in which a group of enzymes called the pyruvate dehydrogenase complex take part (dehydrogenase enzymes remove hydrogen atoms). This complex appears to consist of three different enzymes and five different co-enzymes. A summary of this reaction is

pyruvate + CoA − thioester linkage SH (co-enzyme A) + NAD$^+$ $\longrightarrow$
$\longrightarrow$ acetyl-CoA + CO_2 + NADH + H+

84

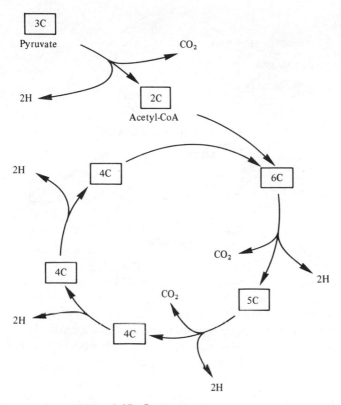

Figure 6.19 *Summary of the TCA cycle.*

The overall reaction is exergonic ($\Delta G^{\ominus}$ = -33.44 kJ mol^{-1}) and in this particular reaction some of the free energy is conserved as the high-energy thioester linkage (-SH) associated with the acetyl-CoA ($\Delta G^{\ominus}$ = -31.35 kJ mol^{-1}). This thioester linkage is used to bind the acetate to the rest of the molecule. Thus co-enzyme A is a higher-energy carrier of acetyl groups. Acetyl-CoA is a high-energy compound, in a similar way in which ATP is a carrier of phosphate groups. In this activated form the acetate is made to enter the TCA cycle.

In the first step of the cycle the acetyl group of acetyl-CoA is transferred to the four-carbon oxaloacetate to form the six-carbon citrate. This is via a six-carbon enzyme-bound intermediate, citroyl-CoA. By this means, free co-enzyme A is formed again (see figure 6.20).

In the reaction that follows, citrate is changed to isocitrate. This change in the form of the molecule converts citrate, which is not readily oxidised,

85

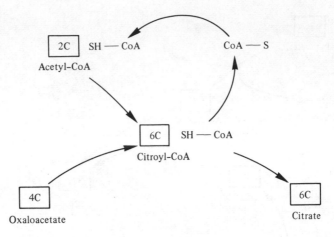

Figure 6.20

into isocitrate which is more readily oxidised. The manoeuvre is achieved by a dehydration reaction (removal of H_2O from the molecule), followed by a rehydration reaction (addition of H_2O). The overall effect of this is to move the hydroxyl group from an internal carbon atom in the molecule to another carbon atom, thus making the molecule more readily oxidisable (figure 6.21).

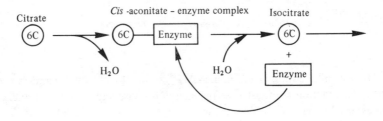

Figure 6.21

The immediately subsequent reactions may be summarised as shown in figure 6.22. Here, the dehydrogenase enzyme involved in the dehydrogenation of isocitrate works only with the NAD present and the reaction takes place with a large free-energy decrease. The two hydrogen ions are therefore accepted by the oxidised form of NAD^+. The following reaction involving oxalosuccinate is also exergonic. Throughout this part of the reaction the oxalosuccinate appears to be bound to the enzyme surface and it is rapidly decarboxylated (CO_2 removed). This part of the sequence can now be looked at in this way. There are two successive oxidation reactions. Firstly oxalosuccinate is formed by oxidation involving the removal of electrons ($NAD^+ \longrightarrow NADH$) and this is then rapidly decarboxylated to produce the five-carbon compound α-ketoglutarate (more recently called 2-oxoglutarate). These oxidative reactions are thermodynamically favourable ($\Delta G^{\ominus} = -20.9$ kJ mol^{-1}).

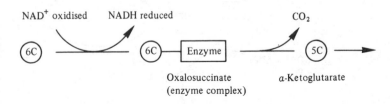

Figure 6.22

The α-ketoglutarate then takes part in a further decarboxylation reaction which in many respects is similar to that which occurs to pyruvate at the beginning of the cycle. A complex of enzymes and co-enzymes is involved and again some of the energy of oxidation is conserved in the thioester group associated with the co-enzyme A; in this case the co-enzyme is succinyl-CoA, rather than acetyl-CoA.

$$\alpha\text{-ketoglutarate} + \text{CoA-SH} + NAD^+ \rightarrow \text{succinyl CoA} + CO_2 + NADH + H^+$$

The oxidation of the α-ketoglutarate results in sufficient free-energy change for the reduction of NAD^+ and the formation of the thioester bond to take place ($\Delta G^{\ominus} = -33.44$ kJ mol^{-1}). Hydrolysis of the resulting thioester bond is then coupled to ATP synthesis (figure 6.23).

thioester bond
succinyl–CoA + guanosine diphosphate + P_i (or ADP in plants)
$\rightarrow$ succinate + CoA + guanosine triphosphate (ATP in plants)

The reaction is only slightly exergonic ($\Delta G^{\ominus} = -2.9$ kJ mol^{-1}) and it can be relatively easily reversed. In mammals a separate enzyme is responsible for conversion of guanosine triphosphate (GTP) to ATP (see chapter 8, Biosynthesis for further explanation of GTP).
Thus

$$\text{succinyl–CoA} + GDP + P_i \rightarrow \text{succinate} + CoA + GTP$$
$$GTP + ADP \rightarrow GDP + ATP$$

87

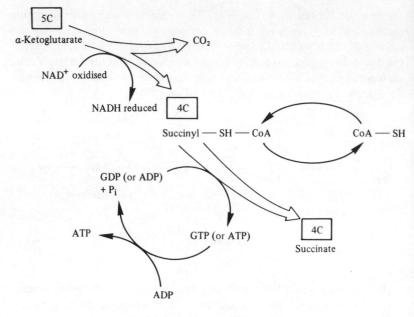

Figure 6.23

This is another example of substrate-level phosphorylation leading to the synthesis of ATP; it is similar to the kind that occurs in glycolysis. It does not involve co-enzyme systems in redox reactions.

In the final part of the cycle the four-carbon compound succinate is converted to the four-carbon compound oxaloacetate. The reactions involve reorganisation of the molecule and, to achieve this, two oxidative steps take place. These changes are brought about by a sequence of three reactions (figure 6.24). Firstly, hydrogen atoms are removed from two of the carbon atoms (C_2 and C_3), and in this way succinate is converted to fumarate. The free-energy change of this reaction is smaller than that of other dehydrogenase reactions in the cycle ($\Delta G^{\ominus} = -219.87$ kJ mol^{-1}). Consequently, the acceptor used here is FAD rather than NAD, with the result that only two ATP molecules are synthesised.

In the second reaction fumarate is converted to malate by a reaction which is more or less reversible ($\Delta G^{\ominus} \approx 0$). In the final reaction, malate is involved in a dehydrogenation reaction involving NAD$^+$, in which oxalo-acetate is formed. NAD$^+$ acts as the electron acceptor but the reaction has a relatively low free-energy change ($\Delta G^{\ominus} = +29.68$ kJ mol^{-1}). The reaction moves in the opposite direction to that indicated by the free-energy

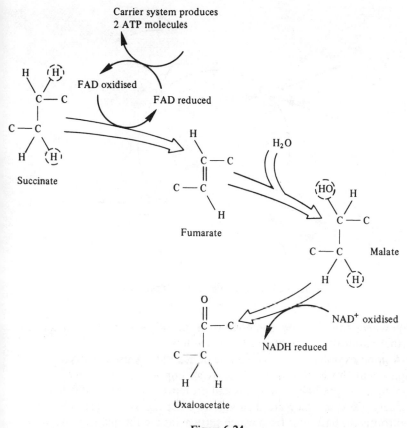

Figure 6.24

change value given, because of the continual utilisation of oxaloacetate in the first stage of the TCA cycle, which is very exergonic.

The purpose of the TCA cycle, as it has been considered so far in this chapter, is the oxidation of acetate to carbon dioxide and the conservation of some of the energy of oxidation as ATP. However, in the living cell there is considerable fluctuation in the level of intermediate compounds within the cycle. An important contributor to this fluctuation of intermediates in the cycle is that these same intermediate compounds are also involved in side reactions. In these reactions the intermediates either form the basis for synthesis of compounds in other pathways (see chapter 8) or else they are increased as a result of other side reactions which produce intermediates for the TCA cycle (figure 6.25). For this reason the TCA cycle is referred to as an **amphibolic** pathway because it functions both in a catabolic way, where oxidation occurs in respiration, and in an anabolic

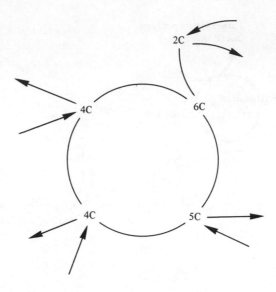

Figure 6.25 *Amphibolic nature of the TCA cycle.*

way where intermediate compounds in the cycle act as precursors for further pathways, many of which are synthetic.

A good example of this is the role of acetyl-CoA itself. Fatty acids and amino acids both yield acetyl-CoA as an important breakdown product. Consequently, carbohydrates, fats and proteins can all be finally converted to acetyl-CoA and they are then able to enter the TCA cycle. Acetyl-CoA also forms the basic unit from which fatty acids are formed (see chapter 8).

The importance of the TCA cycle in catabolism of fats and proteins is obvious when it is seen that acetyl-CoA occupies a significant position in all of these pathways. The decarboxylation of pyruvate is only one source of acetyl-CoA; other sources are the oxidation of fatty acids and the catabolism of amino acids. In this way the TCA cycle is central to the whole of aerobic metabolism.

EVIDENCE FOR THE TCA CYCLE PATHWAY

The process of scientific investigation is through carefully devised experiments which test a particular hypothesis related to the general theory under consideration. From the findings, workers evaluate the evidence to decide whether the hypothesis is supported or the theory requires modification.

The experiments and discoveries that led to the formulation of the TCA cycle by Hans Krebs in 1937 have attained classic status and they are worth brief consideration here. They are a good example of systematic, scientific investigatory work where evidence was either used to support the hypothesis or new interpretations were required. Already, before Krebs' work the Hungarian, Szent-Györgyi, had worked out part of the sequence as a linear pathway

$$succinate \rightarrow fumarate \rightarrow malate \rightarrow oxaloacetate$$

Also the German workers Martius and Knoop had suggested another portion of the pathway

$$citrate \rightarrow \alpha\text{-ketoglutarate} \rightarrow succinate$$

Following on from this initial work, Krebs set about identifying the complete sequence. He chose to use the flight muscles of pigeons because these have a high level of respiratory activity and therefore they seemed to be particularly suited to this type of investigation.

Krebs found that homogenates of the muscles caused rapid oxidation of certain organic acids. All of these acids had been cited by previous workers as part of the pathway. He experimented with other organic acids which previously had not been associated with the pathway, and found that they were not oxidised. If pyruvate was added to those muscle homogenates, then the other organic acids in the pathway seemed to have the effect of increasing the oxidation of the pyruvate. Also when muscle homogenate was incubated under anaerobic conditions with additional oxaloacetate, then citrate was formed. This led him to put forward the idea that what had been considered previously as a linear sequence was better explained if the pathway was a cycle. He had demonstrated a mechanism by which oxaloacetic acid and citric acid were linked (figure 6.25).

Some of Krebs' evidence for the TCA pathway was obtained using malonate, an organic acid which acts as an enzyme inhibitor by specifically blocking the action of succinic dehydrogenase, the enzyme which converts succinate to fumarate in the cycle. By using malonate, he was able to show that the oxidation of pyruvate, which is normally accelerated by the presence of any of the organic acids in the cycle, was actually prevented in all cases. This suggested to him that the oxidation of succinate to fumarate was an essential step in a sequence of reactions in which all of the organic acids identified were involved. Furthermore, he found that there was also a corresponding increase in the accumulation of succinate when added amounts of any of the organic acids such as citrate, isocitrate or α-ketoglutarate were introduced. Also, and crucial to the hypothesis that the pathway is cyclic, he showed that with the addition of malonate there was accumulation of succinate when extra fumarate, malate or oxaloacetate were added (figure 6.26).

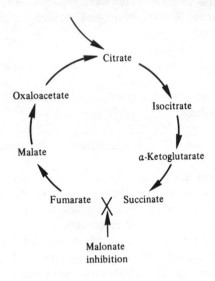

Figure 6.26

Malonate was found to block pyruvate utilisation; however, if the level of oxaloacetate was artificially raised, Krebs demonstrated that pyruvate was again utilised and the rate was proportional to the amount of oxaloacetate added. At the same time, succinate also accumulated. His explanation for this was based on the idea that in the normal situation the oxaloacetic acid is continually regenerated by the cycle, but this ceases when malonate is introduced into the system, and that for each molecule of pyruvate used, one molecule of oxaloacetate is required (figure 6.27).

MITOCHONDRIA, THE TCA CYCLE AND EXCHANGES WITH THE REST OF THE CELL

The linking of the location of the process of the TCA cycle to the mitochondria was made by Kennedy and Lehninger in 1948. These workers showed that isolated mitochondria from homogenised liver of rats could oxidise pyruvate in the presence of oxygen. Furthermore, the rate of

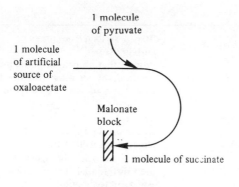

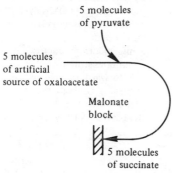

Figure 6.27

pyruvate utilisation and oxygen consumption by the mitochondria was directly related to the rate of respiration in the whole cells. Also, other fractions of the homogenate, apart from the mitochondria, were unable to act in this way.

The study of the role of the mitochondria in the respiratory process has a long and fascinating history. Table 6.1 highlights some of the significant milestones.

Mitochondria are present in almost all aerobic cells of eukaryotes and contain all of the enzymes needed for the TCA cycle. The number of mitochondria per cell seems to be fairly constant for any particular cell type — each liver cell, for instance, contains about 800; but there is considerable variation between the number of mitochondria in different types of cell.

Each mitochondrion is based on a double-membrane structure where the inner membrane is highly folded. The folds are referred to as cristae (figure 6.28). It is on this folded inner membrane surface that electron transport and oxidative phosphorylation take place. All respiratory

93

Table 6.1

1850	Kölliker	first described the presence of mitochondria in muscle tissue.
1900	Michaelis	showed that mitochondria are involved in oxidative reactions by staining with the dye Janus green B, which lost its colour as oxygen consumption proceeded.
1913	Warburg	demonstrated that the total rate of cellular oxygen consumption was associated with the mitochondria.
1937	Krebs	elucidated the TCA cycle.
1940	Claude	isolated mitochondria from liver cells.
1948	Hogeboom, Schneider and Palade	used differential centrifugation of homogenates to isolate mitochondria
1948	Kennedy and Lehninger	showed that the TCA cycle, electron transport and oxidative phosphorylation are carried out in the mitochondria.
1952	Palade and Sjostrand	using electron microscopy showed the double membrane structure.
1961	Mitchell	put forward the chemiosmotic theory of proton extrusion through the inner mitochondrial membrane which together with electron transfer forms the basis of ATP synthesis (see chapter 5).

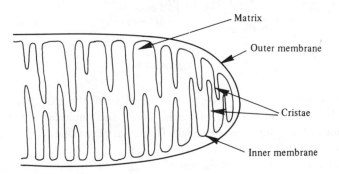

Matrix

Outer membrane

Cristae

Inner membrane

Figure 6.28 *Part of a mitochondrion.*

metabolism beyond pyruvate occurs in the mitochondria and this means that they are the site where most ATP production takes place. As described earlier (chapter 3), the electron carriers of the respiratory chain are integral to the inner membrane and they are arranged in particular sequences, as are ATP synthetase and the proton flux through the ATP synthetase which drives the phosphorylation of ATP from ADP. The binding sites for NADH and oxygen are also on the inner side of the membrane.

The outer membrane of the mitochondrion is relatively permeable but the inner membrane is not and depends on specific carriers for its action. Pyruvate generated by the glycolytic pathway is transported across the membrane, but NAD^+ and NADH are not. Consequently the NADH generated in the cytoplasm must pass electrons inwards to the electron transport chain located on the inner mitochondrial membrane without movement of the co-enzyme itself. All of the NADH formed outside in the cytoplasm must pass electrons to the electron transport chain in the inner membrane without actual movement of the co-enzyme across the membrane. A number of shuttles seem to be able to carry out this function. The principle appears to be the same in each case. The NADH in the cytoplasm passes its electrons to an organic molecule which is able to carry the electrons into the inner membrane of the mitochondrion.

Once there, re-oxidation of the shuttle occurs so that the electrons are passed on to a co-enzyme on the inside of the mitochondrial membrane. A common carrier system across the outer membrane seems to be one that consists of dihydroxyacetone phosphate and glycerol-3-phosphate (see figure 6.29). The glycerol-3-phosphate diffuses through the membrane and is re-oxidised by the dehydrogenase linked to FAD on the outside of the inner membrane. In this particular sequence, NADH passes the electrons to the chain of electron carriers via FAD and the dihydroxyacetone phosphate is able to return to the outside cytoplasm where it can then be re used as a carrier again.

Since FAD passes electrons directly to co-enzyme Q, this sequence does not include the first energy-conserving stage in the normal chain. This therefore yields only two ATP molecules per pair of electrons. Thus the

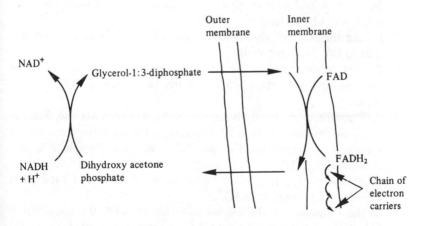

Figure 6.29

ATP yield of NAD-related oxidations is two for cytoplasmic dehydrogenases and three for mitochondrial dehydrogenases. The difference in energy yield between these two systems is related to the energy needed to drive the system which brings electrons into the mitochondrion. Consequently, since there are two molecules of NADH per molecule of glucose used, the number of molecules of ATP produced in aerobic systems is 36 or 38 for eukaryotic cells, depending on the type of shuttle used, and 38 for prokaryotes.

RESPIRATION – SUMMARY

1. Respiration involves catabolism of fuel molecules and the conservation of chemical energy as ATP.
2. The biochemical pathway consists of two main parts – anaerobic glycolysis and then either aerobic catabolism in a cyclic process or one of a number of different anaerobic sequences.
3. Glycolysis occurs in the cytosol. It is not linked to the membrane structures and it takes place in all living things.
4. The aerobic process results in the greatest energy conservation. Energy for ATP synthesis comes from a series of redox reactions in the TCA cycle; this is not the case in glycolysis where there is relatively little overall free-energy change.
5. In glycolysis the intermediate compounds are phosphorylated and these phosphate groups ultimately become the terminal phosphate groups of ATP.
6. In the first reactions of glycolysis, glucose is phosphorylated twice using two ATP molecules.
7. During the stage of glycolysis the 6-C phosphorylated compound is split to two 3-C compounds.
8. Two molecules of ATP are formed from each molecule of glucose used, giving a net overall product of two molecules of ATP for glycolysis.
9. ATP synthesis in glycolysis involves reduction of NAD and then the oxidative reaction where ATP is formed is linked to the formation of a phosphoanhydride bond in the 3-C compound. Phosphate is then transferred to ATP.
10. Oxidation in glycolysis is by the addition of water and subsequent removal of hydrogen by NAD^+.
11. In the formation of the second molecule of ATP, the phosphate is linked by an ester bond to the 3-C compound and this is converted to a phosphoenol bond.

12. This forms phosphoenol pyruvate which occurs in several steps and a compound with a large negative free-energy change on hydrolysis is formed.
13. In anaerobic systems the electrons are passed via NAD^+ and used to reduce the pyruvate formed at the end of glycolysis, thus forming lactate or ethanol or some other product.
14. Often, the starting point for glycolysis is glycogen or starch. These polymers produce greater free-energy changes than the equivalent amount of glucose when catabolism occurs. Thus there is an extra yield of ATP per glucose unit.
15. In aerobic systems the carbon of the pyruvate is oxidised in a cyclic pathway in which the first step is the formation of the 2-C compound, acetyl-CoA.
16. The acetyl-CoA is added to the 4-C oxaloacetate to form citrate, a 6-C compound.
17. Citrate is broken down first to a 5-C and then a 4-C compound. The 4-C compound then has a 2-C compound added to it to make a 6-C compound and thus complete the cycle.
18. Conversion of pyruvate to acetyl-CoA involves the pyruvate dehydrogenase complex. Some of the free energy is conserved as a high-energy thioester linkage which binds the acetyl-CoA to the rest of the molecule.
19. The dehydrogenase reaction involves removal of electrons and protons to NAD^+, except in the case of succinate conversion to fumarate, where the free-energy change is less than in other cases and here the acceptor is FAD.
20. The TCA cycle is an amphibolic pathway involving both the catabolic process of respiratory oxidation and anabolism, where intermediate compounds of the cycle act as precursors for further pathways.
21. Evidence for the actual pathway of the TCA cycle involved the following
    (i) Before Krebs, two parts of the pathway were worked out as linear sequences.
    (ii) Krebs' studies showed oxidation of only certain organic acids in respiring tissues.
    (iii) Addition of pyruvate to respiring tissues caused other organic acids in the pathway to increase.
    (iv) When oxaloacetate was added, then the amount of citrate increased, thus indicating that this is a cyclic process.
    (v) Addition of the specific blocking agent malonate prevented the oxidation of pyruvate, and succinate increased.
    (vi) If malonate was used, but the amount of oxaloacetate was

increased, then pyruvate was utilised and the rate was proportional to the amount of oxaloacetate added. Succinate also accumulated.

7 Photosynthesis

This chapter is concerned with the way in which biological systems have been able to adapt their specific need to synthesise organic materials to the physical properties relating to light, which is a form of energy. Again, some detailed biochemical information is given to show something of the nature of the complexity of the chemical manoeuvres which allow organisms to utilise light energy. Here, as well, investigators have been particularly ingenious in their experiments to test the various hypotheses put forward. What you are presented with here is an outline of a working model for photosynthesis and from time to time some of the evidence for this model is given. The model is still incomplete and no doubt aspects will be changed as further evidence is obtained.

So far we have considered respiration which involves energy metabolism where relatively complex molecules are used as the basic fuel. However, sustainable life depends on a mechanism for the initial synthesis of these complex fuel molecules. This is achieved through photosynthesis, a process which enables an input of energy to be made so that **synthesis** (anabolism) can take place.

Green plants have the special ability of being able to absorb light energy and convert it to chemical energy which is then used in a reduction reaction where carbon dioxide forms glucose (figure 7.1). The processes involved are less well understood than respiration but many of the principles already

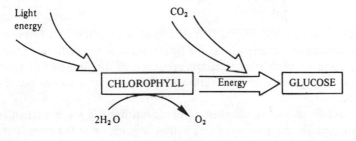

Figure 7.1

discussed in relation to respiratory mechanisms are involved. The overall reaction requires a free-energy input and for higher plants this may be summarised as follows

$$6CO_2 + 6H_2O \rightarrow C_6H_{12}O_6 + 6O_2 \qquad \Delta G^{\ominus} = 2867.5 \text{ kJ mol}^{-1}$$

Other forms of photosynthesis involving different chemical materials may take place in such organisms as photosynthetic bacteria. In the reaction given above, water acts as the reducing agent. Also, it appears that this reaction is the reverse of respiration. Further than that, in respiration, the hydrogen is released from substrates so that the electrons combine with oxygen to form water, the electrons being passed via a chain of carriers which have the ability to couple free-energy changes with ATP formation. In photosynthesis, the electrons removed from water are passed up an 'energy gradient' to be combined with carbon dioxide and this is made possible by an imput of energy (figure 7.2).

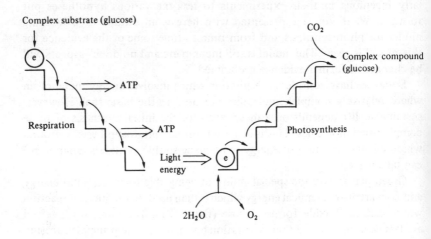

Figure 7.2

Light is needed only for the first part of photosynthesis; for this reason these reactions are referred to as the **light reactions**. Formation of glucose from carbon dioxide using materials produced in the first part of photosynthesis can proceed in the dark and therefore it is called the **dark reaction**.

The study of the mechanism of photosynthesis is made particularly difficult because the process of respiration is going on at the same time in the cells. A breakthrough came in the early 1950s when Arnon and co-

100

workers developed techniques for studying isolated chloroplasts which were still capable of photosynthesis.

The view that oxygen evolved by photosynthesis is derived from water, is supported by elegant experiments carried out as early as 1941 by Rubin, Randell, Kumen and Hyde. They used cultures of the alga, *Chlorella*, in water, labelled with ^{18}O ($H_2^{18}O$), and enriched with carbonate and hydrogencarbonate ions. The normal level of ^{18}O naturally occurring in water is about 0.2 per cent, but in these experiments it was increased to 0.85 per cent. Examination of the oxygen produced by the photosynthesising algae revealed a higher level of ^{18}O than the normal 0.2 per cent. As the experiment proceeded, so the labelled oxygen in the water was exchanged gradually with the normal oxygen isotope. The proportion of labelled ^{18}O remained roughly constant at about 0.8 per cent as the volume increased and as the experiment progressed. These results indicated that the oxygen evolved was derived from the water. Then when further tests were made, using water with the normal 0.2 per cent level of $H_2^{18}O$, and with labelled ^{18}O in carbonate and hydrogencarbonate ions, the proportion of ^{18}O evolved in the oxygen produced by the algae remained at about 0.2 per cent. This again was further evidence that the source of oxygen was not from the carbonate or hydrogencarbonate ions and consequently the oxygen from photosynthesis is generally considered to be derived from water (figure 7.3). Carbonate and hydrogencarbonate ions were used in these experiments, because in solution, carbon dioxide would form these ions and *Chlorella* lives in an aquatic environment.

SOME IMPORTANT PROPERTIES OF LIGHT

Light is a form of **electromagnetic radiation**, being only part of a much longer continuum, the rest of which is not visible (figure 7.4). Planck suggested that light is produced when atoms give off discrete amounts of energy. Current theory is that light may be considered to behave as though it consists of waves of particles; each particle is called a **photon**. Thus each photon is actually a unit of light energy. According to the quantum theory of physics, energy is emitted from bodies that radiate energy and this energy is released as separate units; each separate unit is referred to as a **quantum** of energy. Consequently a quantum of light energy is a photon.

The **intensity** of light is determined by the number of photons emitted each second. The **wavelength** of the light, sometimes called the **frequency**, is decided by the particular energy level of the photons being emitted. Therefore light of different wavelengths is a different colour and has different amounts of energy per photon. Thus white light consists of

101

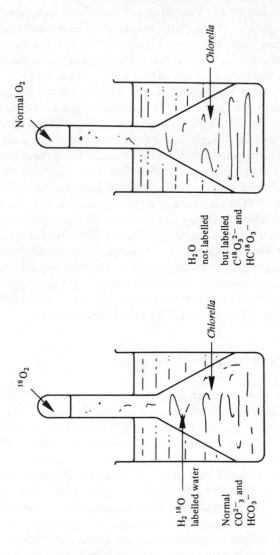

Figure 7.3

Normal O$_2$

^{18}O$_2$

Chlorella

Chlorella

H$_2$O
not labelled
but labelled
C^{18}O$_3^{2-}$ and
HC^{18}O$_3^-$

H$_2$^{18}O
labelled water

Normal
CO$_3^{2-}$ and
HCO$_3^-$

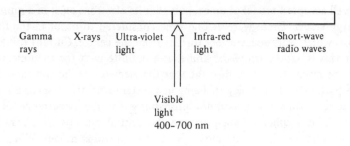

| Gamma rays | X-rays | Ultra-violet light | Infra-red light | Short-wave radio waves |

Visible
light
400–700 nm

Figure 7.4

photons of many different energy levels. Any particular wavelength of light is related to the number of oscillations made each second (frequency); the more oscillations that there are, then the shorter the distance between each oscillation and the shorter the wavelength. For instance, blue light has a relatively short wavelength ($\approx$ 450 nm), and red light has a much longer wavelength ($\approx$ 700 nm) (figure 7.5).

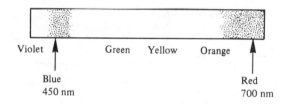

| Violet | Green | Yellow | Orange |

Blue
450 nm

Red
700 nm

Figure 7.5 *Relationship between wavelength of oscillations and colour of light.*

THE PHOTOELECTRIC EFFECT AND ENERGY LEVELS OF ELECTRONS

A phenomenon that has been known for some time is that light is capable of causing a **photoelectric effect** when absorbed by certain types of substances. When this happens an electric charge is set up which will flow in a wire.

An explanation for this is that the light photons are absorbed by the surface layer of atoms and their energy is transformed into the kinetic energy of the electrons. When the photoelectric effect occurs, the photon delivers its entire energy to a single electron within a surface molecule of the material in which it is absorbed.

103

A well-accepted model of an atom is that its electrons exist in a number of energy states. These were at one time considered to occur in the form of orbits which were successively further from the centre. Current thinking is that this is rather simplistic and that a definite path for each electron cannot be given. Now it is thought that the electron can be considered to have a particular probability of being in a certain part of the space around the nucleus. Thus the maximum probability for the occurrence of an electron is in a spherical shell around the central nucleus with a radius equal to that of the orbits envisaged in the previous model. When the electron occupies a shell nearest to the central nucleus, this is referred to as the **ground state** and it is at its lowest energy level. However, the electron can be made to move out to one of the shells further from the centre where it has a higher energy level. Such an electron is now said to be in an **excited state**; theoretically there are a number of energy levels that the electron can move out to. If an electron acquires sufficient energy, and this can happen when it collides with a photon, it may be removed altogether from its shell around the nucleus. Electrons may be moved to higher energy levels or removed entirely from around the nucleus by photons—when light is absorbed. However, when electrons move to lower energy levels, light is emitted (figure 7.6).

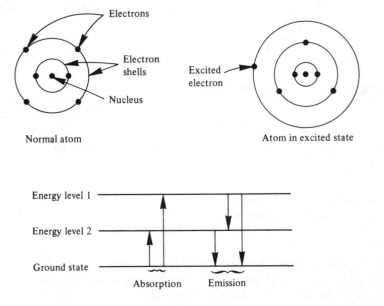

Figure 7.6

In this way, light can cause an electron to move from one shell to another; however, the difference between the energy levels must be the same as the energy of the photon that impinges upon it, otherwise the excitation will not occur. Molecules, therefore, absorb light selectively from particular wavelengths. The same applies to the photoelectric effect; again, the wavelength and molecules within the material must both be appropriate otherwise it will not occur. The photoelectric effect involves electrons being removed entirely from shells around the nucleus. These electrons are now free and form a flow of **electric charge**.

Electrons in the excited state cannot remain in this condition for long; they may return to the ground state directly. The excess energy may be released in the form of a photon of light and then the substance shows fluorescence; some of the energy is emitted as heat in accordance with thermodynamic laws (see chapter 1). Excited electrons can also return to the ground state by **transfer** of energy to electrons in a neighbouring molecule. This is referred to as **resonance energy**; in other cases, the electron itself may be transferred to an adjacent molecule. As will be seen later, these latter two mechanisms are important in photosynthesis.

The redox potential (E_m) for removal of an electron from an excited molecule is more negative than the E_m for the molecule in the ground state. This E_m value can be expressed in electron volts (eV).

CHLOROPHYLL

As mentioned previously, the ability to absorb light varies from one substance to another and different substances are able to absorb different wavelengths to a greater or smaller extent. Very few substances of biological importance are altered chemically by visible radiation, but chlorophyll is an exception. Chlorophyll absorbs light of wavelengths in the blue/violet and red parts of the spectrum, but green light is transmitted; consequently the chlorophyll-bearing parts of plants appear green. A phenomenon similar to the photoelectric effect occurs in chlorophyll, but here, instead of the electrons which leave the shell of the parent nucleus creating a flow of electric charge, the electrons are taken up by a series of electron carriers similar to those used in ATP synthesis in the respiratory pathway (see chapter 6).

There are two common types of chlorophyll found in higher plants; they are called chlorophyll a and b, and both are important in photosynthesis. Chemically they are closely related and those who wish to know more about the chemical structure of chlorophyll, and the differences between the types of chlorophyll, are advised to refer to advanced texts which usually give this type of information (for example, see the book by R. K. Clayton — details in the Bibliography at the end of this book).

Basically, the mode of operation of chlorophyll seems to be that in the ground state, the reduction potential is much lower than in the excited state. The effect of this is that chlorophyll in the excited state becomes a strong reducing agent. The chlorophyll therefore releases an electron to become oxidised and, at the same time, another molecule which is the receptor of the electron is reduced. The chlorophyll has now been converted to a strong oxidant so that it is able to receive an electron from a third type of molecule which acts as a donor. In this way light initiates the formation of reducing agents and oxidising agents (figure 7.7).

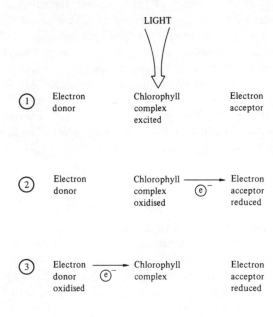

Figure 7.7

The reducing potential created by the effect of light on chlorophyll is later used to reduce carbon dioxide to form carbohydrates. The consequent oxidising power is used to oxidise water by removal of hydrogen to form oxygen in higher plants (figure 7.8).

The light-induced oxidation of substances like chlorophylls can be detected by measuring changes in those parts of the visible spectrum that are absorbed. With oxidation, the well-defined absorption band in the red

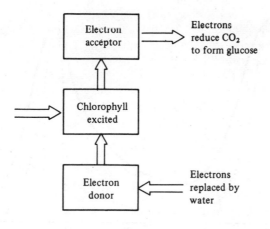

Figure 7.8

part of the spectrum is lost. Observations have shown that there are two main absorption bands for the oxidised form of chlorophyll. These are at 700 nm and 680 nm and the differences are significant. One is referred to as **photosystem I** (P_I or P_{700}) and the other **photosystem II** (P_{II} or P_{680}). As far as has been made out so far, both P_{700} and P_{680} are chlorophyll a molecules attached to different specific proteins.

As early as the 1930s, Hill discovered that isolated chloroplasts when illuminated are capable of reducing certain electron acceptors such as dyes. These dyes change colour from the oxidised to the reduced state. Hill found that when the dyes were reduced, oxygen was produced. This led to the idea that electron acceptors probably occur in chloroplasts. From this, nicotinamide adenine dinucleotide (NADP) was identified; as will be seen later, this has a central role in photosynthesis.

Part of the evidence that chlorophyll is the main light-absorbing agent for photosynthesis, is the close correspondence between the **absorption spectra** of chlorophyll and the **efficiency spectra** of light in photosynthesis (figure 7.9).

THE STRUCTURE OF CHLOROPLASTS

The light-dependent reactions and the dark reactions of photosynthesis in higher plants all take place in the chloroplasts. Photosynthetic bacteria are different since the photochemical reactions take place in the cell membrane and there are no chloroplasts. In higher plants, the chloroplasts may be of different shapes in different species, but they are usually ellip-

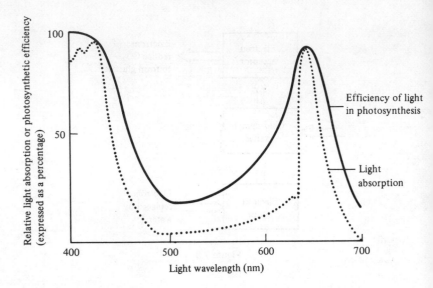

Figure 7.9 *Graphs to show the absorption spectrum of chlorophyll a and the efficiency of light of different wavelengths in photosynthesis.*

soidal and occur in large numbers in each cell. Individual chloroplasts are bounded by a double membrane, with the inner one highly folded, not unlike the cristae of mitochondria (figure 7.10). The membranous folds are called the **lamellae** and at intervals they form flattened disc-like structures which are the **thylakoids**. In certain areas the thylakoids are stacked one on the other, rather like a pile of coins. These structures are the **grana** and it is here that the pigments, electron carriers and ATPase enzymes involved in the light-dependent reactions are located (figure 7.11). Most of the

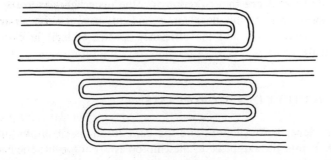

Figure 7.10 *Folding of the inner membrane of the chloroplast, to show stacking to form grana.*

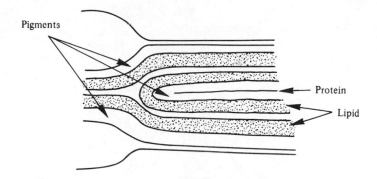

Pigments

Protein

Lipid

Figure 7.11 *Detail of part of a granum to show arrangement of layers.*

enzymes of the dark reaction are in the **stroma**, which is the fluid medium surrounding the thylakoids.

PHOTOSYNTHETIC UNITS

One current theory is that chloroplasts in higher plants contain layers of proteins and various pigments, particularly chlorophylls and carotenoids, and that these are grouped into photosynthetic units each consisting of about 300–400 molecules of pigment. The way in which they function is that light impinges on the pigment molecules and this initiates a chain reaction, probably based on transmission by resonance (see page 105). The presence of a variety of pigments will extend the range of wavelengths that are available for photosynthesis. Each photosynthetic unit contains central chlorophyll molecules which are either P_{700} or P_{680} and this forms what is known as the **reaction centre**. Here, light-induced oxidation occurs (figure 7.12).

The most purified reaction centres involving photosystem I appear to contain about 40 molecules of chlorophyll a and only part of this is capable of chemical activity as P_{700}. Similarly in photosystem II there seem to be about 40 molecules of chlorophyll a and a small part exists as P_{680}. Only a very small proportion of the total pigment in chloroplasts is made up of chlorophyll molecules of P_{700} or P_{680}. Thus most of the pigment is not photochemically active but seems to have the role of absorbing photons and passing this energy on to the photochemically active part of the reaction centre. These other pigment molecules are referred to as '**antenna**' molecules and the energy is passed to the P_{700}

109

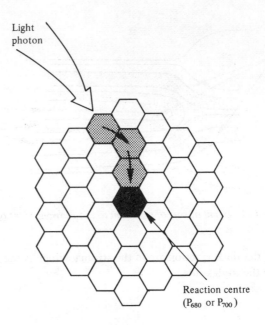

Light
photon

Reaction centre
(P_{680} or P_{700})

Figure 7.12 *Diagram of a photosynthetic unit to show the position of a reaction centre.*

or P_{680} at the centre. By having antenna systems the efficiency of collecting photons is considerably enhanced.

THE LIGHT REACTIONS

There appear to be two types of light-dependent reactions that occur in chloroplasts. One is a **non-cyclic electron transport pathway** and it results in the accumulation of a reduced product (NADPH) and synthesis of ATP. The other is a **cyclic electron flow** and here electrons leave the chlorophyll molecule when illuminated and pass along a chain of electron carriers to be returned to chlorophyll. In this way there is no accumulation of reduced product but ATP is formed. Some of the same carriers are used in both pathways. It is not clear how these two pathways are linked but together they result in the formation of ATP and NADPH which are required for the synthesis of glucose from CO_2.

In 1954 Arnon and Whatley provided important evidence by demonstrating that isolated chloroplasts can produce ATP from ADP and inorganic phosphate in the light. Arnon's suggestion is that there are two

110

distinct photochemical processes, cyclic and non-cyclic photophosphorylation acting as though they are in parallel rather than in series.

The photochemical reactions of photosynthetic bacteria are similar to
those of higher plants; however, no oxygen is evolved and only one photosystem seems to be used. Investigation of bacterial photosynthetic pathways has proved valuable since they are easier to study than higher plants
because of the lack of chloroplasts, but there are dangers when using
parallel or similar systems and investigators have had to be alert to the
potential dangers of assuming too much similarity.

NON-CYCLIC PHOTOPHOSPHORYLATION

In higher plants, two photochemical reactions are necessary to transfer
electrons from water to $NADP^+$. These reactions are brought about by
photosystems II and I (summarised in figure 7.13). This is sometimes
referred to as the Z scheme. Here photoexcitation of P_{680} produces a

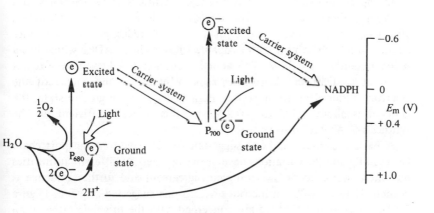

Figure 7.13

strong oxidant which oxidises water by removal of hydrogen to form
oxygen, and the reductant formed releases electrons to a chain of carriers
that links to P_{700}. The resulting excitation of P_{700} creates a strong reducing
agent which passes electrons to $NADP^+$ through a system of carriers. Oxidation of water to oxygen requires removal of four electrons from chlorophyll for each molecule of oxygen produced. The excited electrons lost
from P_{680} are replaced by electrons from hydroxyl ions derived from
water

111

$$H_2O \rightarrow H^+ + OH^-$$
$$OH^- \rightarrow \tfrac{1}{2}O_2 + H^+ + (2e^-) \rightarrow \text{ to } P_{680}$$

The replacement of electrons later lost from P_{700} comes from those lost from P_{680} This system of linking enables the second photosystem in the series to act as an electron booster for the first. There is an input of energy by light into both photosystems I and II. If the system is considered from the point of view of the relative redox potential of the initial reactants compared to the final stage in photosystem I, the electrons are raised from almost 1.0 V to -0.32 V. It has to be remembered that the more negative the redox potential, the greater the tendency to lose electrons. In the first part of the Z system, absorption of light quanta in P_{II} raises electrons from about 1.0 V to 0.1 V. The electrons move through a chain of carriers from excited P_{680} to P_{700} in the ground state. Also there is enough potential difference between water and P_{II} in the ground state for electrons to move to P_{II}.

In the ground state, P_{700} has a redox potential of between 0.4 V and 0.5 V, and therefore it has little tendency to lose an electron. However, after P_{700} has been excited by light quanta, the redox potential becomes about -0.5 V. It is now quite capable of reducing $NADP^+$ which has a redox potential of 0.32 V. The actual reduction is achieved by a redox potential gradient of electron carriers. Also, along the course of the electron carrier chain from excited P_{680} to P_{700} in the ground state, the gradient is about 0.3 V to -0.4 V and this is quite sufficient for the synthesis of ATP from ADP.

A basic sequence has been suggested for the electron transport chain between P_{II} and P_I. Variations have been suggested by different authorities but there seems to be no complete agreement and further evidence is needed. Here we will confine ourselves to one suggested sequence (figure 7.14), remembering that we are concerned with the principle rather than the detail.

The elucidation of this pathway by investigators was based on detailed consideration of a number of characteristics of the compounds involved, such as knowledge of standard redox potentials of each carrier, the action of inhibitors and artificial electron carriers, as well as sequences of oxidation and reduction found by using spectroscopy where changes in absorption bands have been observed. Using this type of information the suggested sequence was pieced together.

The first electron acceptor from P_{II} seems to be pheophytin from P_I. There is some uncertainty about the nature of the initial electron acceptor but it is sometimes referred to as P_{430} since it has characteristic light

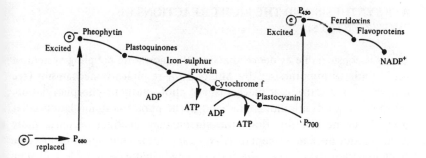

Figure 7.14

absorption changes at wavelengths of 430 nm when it is reduced. It is probably an iron-sulphur protein.

CYCLIC PHOTOPHOSPHORYLATION

It has been shown that isolated chloroplasts can form ATP when illuminated in the absence of added electron donor or acceptor substances. Also there is no accumulation of reduced substrate. From this it appears that light quanta cause electrons to move from excited photosystem I using a circular chain of carriers back to P_I again (figure 7.15). This reaction results in no oxygen being evolved or NADPH production, but there is synthesis of ATP.

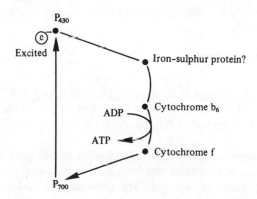

Figure 7.15

ATP SYNTHESIS AND THE LIGHT REACTIONS OF PHOTOSYNTHESIS

There is considerable evidence that the chemiosmotic coupling system of ATP synthesis hypothesised by Mitchell for respiratory mechanisms (see chapter 5) is also the basic mechanism of photosynthetic phosphorylation. When chloroplasts are illuminated and cyclic photophosphorylation takes place involving electron flow, the surrounding medium becomes more alkaline, showing a movement of H^+ ions across the chloroplast membrane. If illumination then ceases, the surrounding medium becomes less alkaline because of the return of H^+ ions. In fact this appears to be in the opposite direction to that which occurs in mitochondria. However the basic principle could be similar, since it is the proton pump which is the important part because it creates a proton gradient which can be used for ATP synthesis rather than the direction of movement. Further evidence for the chemiosmotic system being responsible for ATP synthesis in photosynthesis, is that artificial pH gradients set up across the membrane, even without illumination, result in ATP synthesis.

In non-cyclic photophosphorylation, the movement of electrons from P_{II} to P_I is associated with transport of protons across the thylakoid membrane. Quinones are involved and they have a known part to play in proton movement elsewhere in biological systems.

The significance of the two types of photophosphorylation is not entirely clear; however, the role of cyclic photophosphorylation is possibly to increase the supply of ATP necessary for CO_2 fixation. It is known that the fixation of each molecule of CO_2 involves three molecules of ATP and two molecules of NADPH. It is suggested that the NADPH and two of the ATP molecules are produced by the non-cyclic photophosphorylation and one molecule of ATP comes from the cyclic pathway.

GLUCOSE SYNTHESIS

The formation of glucose is often referred to as a dark reaction; in other words it is not light dependent. The process begins with CO_2 and uses the NADPH and ATP produced by the light reactions to drive the CO_2 fixation (figure 7.16).

The most significant investigations relating to this part of the photosynthetic process were carried out by Calvin and co-workers in the 1950s using radioactive tracers. Again the alga *Chlorella* was used in these experiments. Here the alga was subjected to brief periods of illumination in the presence of $^{14}CO_2$. The algal cells were then immediately killed, by

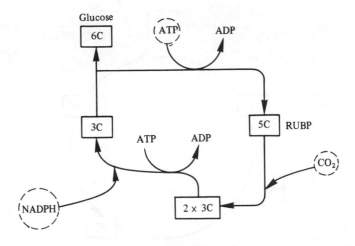

Figure 7.16

dropping them in boiling alcohol, and were then broken up and a search made for organic substances which were labelled with ^{14}C. The method used was a combination of paper chromatography followed by the use of X-ray film where the radioactive material was revealed as dark spots. This technique showed that **3-phosphoglyceric acid (PGA)** was among the compounds first labelled. It is significant that this substance is also a known intermediate in the glycolytic pathway (see chapter 6).

PGA was recognised as a three-carbon compound, the reasoning at the time being that if the carbon from CO_2 forms part of the molecule, then the precursor to PGA should be a two-carbon compound. Such a compound was never found and, ultimately, investigations indicated that the precursor was more likely to be **ribulose-1:5-biphosphate (RUBP)** which is a five-carbon compound. In this model it means that when CO_2 is combined with the RUBP, it is suggested that an unstable six-carbon intermediate compound is formed which splits to form two molecules of PGA. Simply, the pathway would run as shown in figure 7.17. Thus six turns of the pathway produce one molecule of glucose.

Although it is possible to suggest a mechanism for glucose synthesis from two molecules of PGA through a process which is the reverse of the first few reactions of glycolysis, this would not account for the fact that all of the six carbon atoms in the glucose formed from photosynthesis were found to be derived from CO_2.

Calvin managed to piece together a complex cycle, which he suggested more satisfactorily fitted the available evidence; it is now known as the

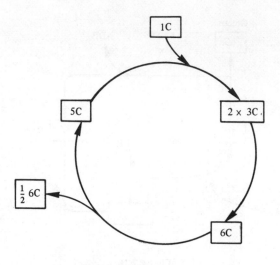

Figure 7.17

Calvin cycle or C_3 pathway. This cycle involves organic compounds with from three to seven carbon atoms in various sugars and their phosphorylated derivatives. In each turn of the cycle, one molecule of RUBP is regenerated for each molecule of CO_2 reduced. The NADPH produced from the non-cyclic photophosphorylation described previously is the reductant for the CO_2. The reactions within the cycle involve interchange between molecules. The suggested overall cycle is outlined in figure 7.18 and some of the steps are given in figure 7.19.

THE C_4 PATHWAY

In the 1960s it was found that 3-PGA may not be the first intermediate compound formed in some plant types such as maize and sugar cane. These are efficient photosynthesisers which grow particularly well under high light intensities. Here a four-carbon compound rather than a three-carbon compound appears to be formed first, and consequently these are known as **C_4 plants**. In this particular pathway the mesophyll cells of the leaf take up carbon dioxide from the air by a process of carboxylation of phosphoenol pyruvate (PEP) to form oxaloacetate. This is then reduced by NADPH to form malate (figure 7.20). Malate is then translocated from the mesophyll cells to the bundle sheath cells around the vascular structures in the interior of the leaf. Here the malate is decarboxylated to pyruvate and

116

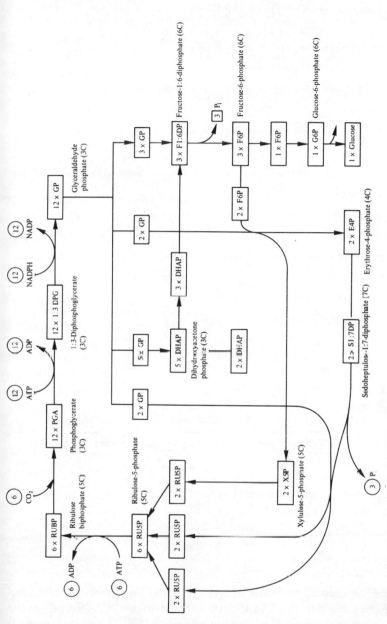

It should be noted that 3 molecules of ATP and 2 molecules of NADPH are required for each molecule of carbon dioxide reduced

117

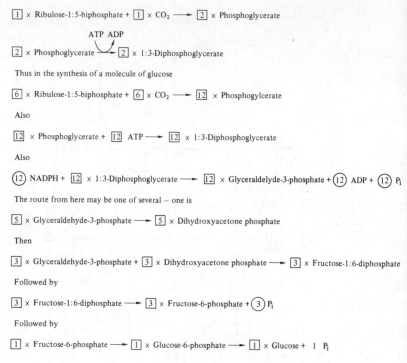

Figure 7.19 *Some important steps in the Calvin cycle.*

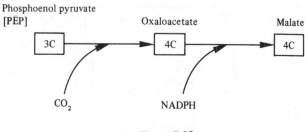

Figure 7.20

$NADP^+$ is reduced to NADPH. The CO_2 from the decarboxylation reaction is passed into the Calvin cycle and the pyruvate is returned to the mesophyll cells where it is phosphorylated to PEP. The whole process is driven by ATP breakdown to AMP; it is a cyclic process and it is referred to as the **Hatch–Slack pathway**.

118

At first sight it may appear unusual that plants fix CO_2 as oxaloacetate in their mesophyll cells only to be decarboxylated and refixed again in another type of cell. However, the significance is that PEP carboxylase has a much higher affinity for CO_2 than RUBP carboxylase, consequently mesophyll cells collect CO_2 with greater efficiency using this mechanism. The effect of this is to produce a high concentration of CO_2 in the bundle sheath cells when malate is decarboxylated (figure 7.21). At the same time it should be appreciated that this pathway requires two extra ATP molecules for each molecule of CO_2 fixed, in other words this means twelve more ATP molecules per molecule of glucose synthesised. Thus C_4 plants tend to grow well only under high light intensities such as in the tropics where they are able to produce the larger amount of ATP required in the

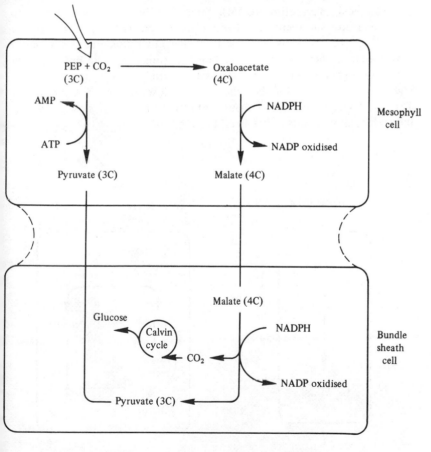

Figure 7.21 *Outline of the Hatch–Slack pathway.*

light reaction. As a result of this, when conditions are suitable, C_4 plants can produce glucose much faster for the same leaf area than C_3 plants.

THE CAM SYSTEM

There is yet a third type of carbon-fixation method in plants. It is rather similar to C_4 metabolism but instead of separating the metabolites between two types of cell, the PEP carboxylase is used to fix CO_2 in malate molecules, which are stored until the next day, when carbon is released and then refixed in the normal Calvin cycle. In this process the interval between day and night is important (figure 7.22). Since the pathway was first discovered in the group of plants called the *Crassulaceae* it was named **crassulacean acid metabolism** (CAM). Subsequently it has been found to occur in many succulent and desert plant types. The importance of the process for plants living in these particular environmental conditions is that it enables them to open their stomata at night to take in CO_2, when humidity is greater, and this reduces transpiration losses. During periods of high light intensity in the day, the stomata close and CO_2 does not enter. This mechanism does not enhance efficiency, in fact it slows it down, but it does allow plants to live in arid regions.

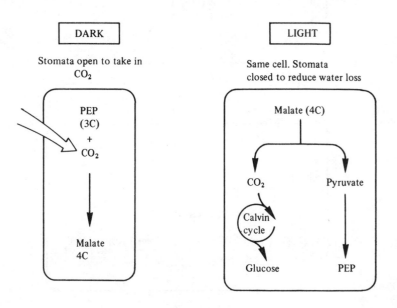

Figure 7.22

PHOTORESPIRATION

As has been mentioned previously (chapter 6), it frequently happens in the study of the biochemical pathways of living things that it is difficult to investigate one process in isolation from all others. This is true in the case of photosynthesis, since the study of photosynthesis has been rendered more difficult because of the phenomenon of **photorespiration**. This reaction occurs in green plant cells and here oxygen takes the place of CO_2 as a substrate. The products formed are a two-carbon compound, phosphoglycolate and a three-carbon compound 3-phosphoglycerate. The phosphoglycolate is oxidised to CO_2 by oxygen in the cytoplasm, mitochondria and organelles called peroxisomes (figure 7.23).

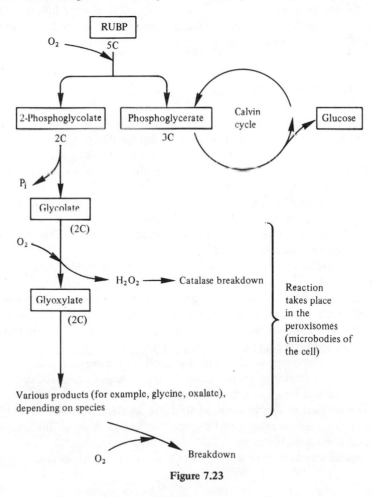

Figure 7.23

121

It appears that as much as one-third of the CO_2 fixed by plants in photosynthesis is released again in photorespiration. The rate of the reaction is increased as light intensity increases. There is no accompanying oxidative phosphorylation to form ATP and the overall significance of the reaction is not at all clear. Photorespiration is active in C_3 plants but insignificant in C_4 plants, thus giving a great advantage to C_4 plants particularly in strongly lit conditions.

Ribulose biphosphate carboxylase is inhibited by molecular oxygen. This appears to be a mechanism for regulation of CO_2 fixation where the partial pressure of oxygen is high and that for CO_2 is low. This enzyme has a much higher affinity for CO_2 than oxygen, but in photosynthesising chloroplasts, the partial pressure of oxygen is maintained at a high level by the light reaction, and the partial pressure of CO_2 is kept low by CO_2 fixation. Under these conditions oxygen becomes a competitor with CO_2 in the RUBP carboxylase reaction.

PHOTOSYNTHESIS – SUMMARY

1. The overall reaction of photosynthesis requires a free-energy input. In higher plants, water acts as the reducing agent.
2. Hydrogen is released from substrates so that the electrons are passed along a chain of carriers which couple the free-energy changes with ATP formation before combination with oxygen to form water.
3. Photosynthesis occurs in two main parts – the light reactions and then carbon fixation which can occur in the dark.
4. Light behaves as though it consists of waves of particles. Each particle is a photon which is a unit of light energy.
5. The wavelength of the light is determined by the energy level of the photons emitted.
6. Certain surfaces are capable of producing a photoelectric effect when light falls on them. Here, photons deliver their energy to electrons in the material of the surface on which they fall. A similar phenomenon occurs with chlorophyll in photosynthesis.
7. In this phenomenon, electrons are caused to move to higher energy levels, thus having an excited state. Electrons may be removed entirely from their orbit around the nucleus.
8. The amount of energy required to move an electron from one energy level to another must equal exactly the energy level of the photon that has caused the move.
9. Excited electrons may return to the ground state by transfer of energy

to electrons in a neighbouring molecule, or the electron itself may be transferred to adjacent molecules by a series of electron carriers.

10. In the excited state, chlorophyll is a strong reducing agent and this releases an electron to become oxidised. At the same time, the receptor of the electron is reduced. Thus chlorophyll is an oxidising agent and a reducing agent.

11. The reducing potential is used to reduce carbon dioxide in the formation of carbohydrates and the oxidising power is used to oxidise water by removal of hydrogen.

12. Two types of chlorophyll are involved in photophosphorylation — P_I and P_{II}, each having different light absorption bands.

13. Chloroplasts have pigments in photosynthetic units of 300–400 molecules.

14. Each unit has a central core of P_I or P_{II} molecules which form the reaction centre. The outer pigment molecules are the 'antenna' molecules which enhance the efficiency of the collection of photons.

15. There are two types of light reaction — cyclic and non-cyclic photophosphorylation.

16. The non-cyclic process results in the accumulation of NADPH and synthesis of ATP.

17. Also P_{II} and P_I are involved in the Z scheme where photoexcitation of P_{II} forms a strong oxidant which oxidises water to oxygen.

18. The reductant formed releases electrons to carriers which link to P_I.

19. P_I creates a strong reducing agent which passes electrons to NADP via a series of carriers.

20. Electrons lost from P_{II} come from water, and electrons lost from P_I go to reduce NADP; these are replaced by electrons from P_{II}.

21. There is an input of energy into both photosystems II and I.

22. As electrons are passed along the chain of carriers, ATP is synthesised.

23. Only P_I is involved in the cyclic process. Here, electrons move via a cyclic path of carriers from P_I and back again. ATP is formed, but not NADPH.

24. There is evidence that ATP synthesis in photosynthesis is also based on a chemiosmotic coupling system.

25. The dark reaction begins with carbon dioxide; NADPH and ATP produced by the light reactions are used to drive the reaction.

26. The carbon fixation pathway involves RUBP (5-C) which is combined with carbon dioxide which is then broken down to two molecules of PGA (3-C).

27. The pathway (Calvin cycle) then involves organic compounds of three-carbon to seven-carbon atoms. Here, one molecule of RUBP is regen-

erated for each molecule of carbon dioxide reduced. NADPH acts as the reductant.

28. For every six molecules of carbon dioxide reduced, one molecule of glucose is formed.

29. Some plants use a different pathway in which a four-carbon compound is an important intermediate. These are called C_4 plants.

30. In the C_4 pathway in the mesophyll cells, carbon dioxide is used to carboxylate PEP (3-C) to form oxaloacetate (4-C).

31. Then oxaloacetate is reduced by NADPH to malate.

32. Malate is translocated to the bundle sheath cells, where pyruvate is formed, and the carbon dioxide released is used in the Calvin cycle.

33. Pyruvate is returned to the mesophyll cells where it is phosphorylated by ATP to PEP.

34. PEP carboxylase has a higher affinity for carbon dioxide than RUBP carboxylase, thus giving more efficient carbon dioxide collection.

35. The C_4 pathway requires more ATP for each molecule of glucose produced. This is derived from the light reactions, thus these plants grow only in high light intensities.

36. In the CAM system the C_4 process occurs in one type of cell but the malate formed is stored overnight. The carbon dioxide released is refixed in the Calvin cycle in the day.

37. In the process of photorespiration, intermediates of the Calvin cycle are broken down to form a variety of products in the presence of oxygen, thus greatly reducing the efficiency of photosynthesis.

8 Biosynthesis

The range of organic materials synthesised within the cells of living organisms is enormous and the pathways involved are often quite bewildering in their complexity. The purpose of this chapter is to try to establish some basic principles linking biosynthesis (anabolism) to the energy metabolism of the cell rather than to attempt any systematic coverage of the different pathways involved. In all cases the processes under discussion have been simplified in order to bring out the principles more clearly. However, readers should appreciate that here, as elsewhere in this book, all processes described are merely the currently accepted working explanation that is broadly agreed by many biochemists and physiologists working in the field. Even here there will often be disagreement among experts about detail and sometimes about significant aspects of the explanation. Also, in this book, it is only simplified explanations of these complex current theories that are under discussion. This simplification in itself can present difficulties, since, by the very process of trying to simplify things, all aspects of the theory cannot be included.

Synthetic reactions within cells later lead to the formation of materials which are characteristic of cells derived from simpler building materials. These materials may then go on to be assembled into more complex arrangements such as membranes or contractile structures and from these the components of cellular organelles are constructed.

As might be expected from synthesis, which is essentially a process where entropy decreases (chapter 1), the reactions are overall endergonic. Such reactions are not spontaneous, therefore they must be coupled to other reactions which are even more strongly exergonic. The method of overcoming this type of problem in biological systems is to couple the hydrolysis of ATP, or other phosphorylated 'high-energy' compounds, to such endergonic reactions (figure 8.1).

Although phosphorylated adenosines such as AMP, ADP and ATP occur in all living cells, there are other phosphate compounds which behave similarly to these adenosine compounds; they are based on other related nucleosides such as guanine, uracil or cytosine. The abbreviations such as

125

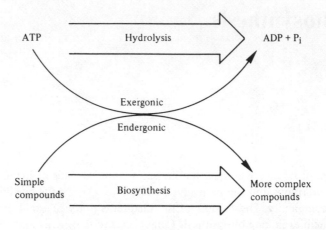

Figure 8.1

GTP or CTP are used. Their function is to transfer phosphate bond energy from ATP to other specific biosynthetic pathways (see chapter 5). Thus

$$ATP + CDP \rightleftharpoons ADP + CTP$$

The CTP is then used in fatty acid synthesis. The GTP produced in other similar reactions would be used for protein synthesis.

For our purposes an outline of the synthesis of the polysaccharide glycogen, fatty acids, amino acids and proteins only will be considered, as from these types of materials most other biologically important compounds can be formed.

The initial starting materials required by cells for synthesis vary, depending on the enzyme systems available. Many heterotrophic cells can produce all of the necessary cell components from glucose, nitrogen (which is provided in the form of ammonia or amino acids), and trace minerals and small quantities of a limited group of organic compounds. The ability to create such a diversity of chemical materials from a limited selection of simple materials is due partly to the fact that the central metabolic pathways are closely interrelated (figure 8.2). We have already considered the amphibolic nature of the TCA cycle (page 89), but the important role of the glycolytic process and TCA cycle in providing not only ATP, but also a supply of reduced NADPH and a range of the required intermediates, should be appreciated.

An important principle of metabolic pathways is that synthesis and breakdown (catabolism) cannot normally use complete sequences which are totally common. For instance, the breakdown of glucose to pyruvate

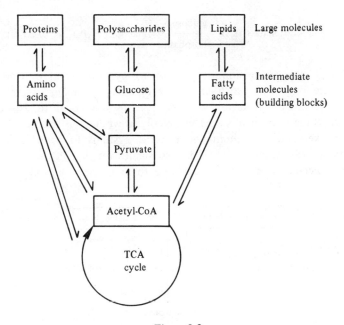

Figure 8.2

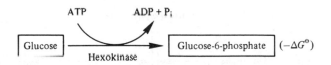

Figure 8.3

in glycolysis cannot be simply reversed to synthesise glucose from pyruvate. Often in a particular catabolic pathway there are individual reactions which involve a relatively large negative free-energy change. This means that such reactions, and consequently the whole pathway, are essentially irreversible. Thus an alternative pathway, or part of the pathway, is necessary if the reverse process is to be accomplished. An example of this is in the reaction where glucose is converted to glucose-6-phosphate in glycolysis; here the enzyme hexokinase is involved and ATP is used to phosphorylate the glucose. The reaction has a negative free-energy change and therefore occurs spontaneously (figure 8.3). Clearly this reaction cannot be readily reversed, therefore in the process where glucose is synthesised from pyruvate, a different enzyme system is employed for the reverse reaction

127

$$\text{glucose-6-phosphate} + H_2O \xrightarrow{\text{glucose-6-phosphatase}} \text{glucose} + H_3PO_4 \quad (-\Delta G^{\ominus})$$

Although the conversion of glucose to pyruvate, and pyruvate to glucose, are pathways which are not reversible, a number of the steps in the sequence are reversible and they can be used in both. However, there are three particular stages in glycolysis where the reaction is sufficiently exergonic for alternative enzyme systems to be used in the synthetic process (figure 8.4). It is also significant that in the glycolysis of one molecule of glucose to pyruvate there is the net production of

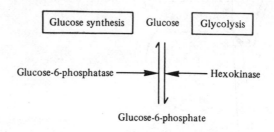

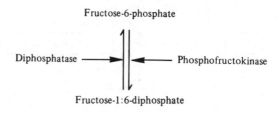

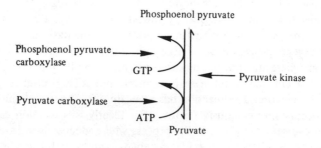

Figure 8.4

two molecules of ATP. The reverse anabolic process requires the input of four molecules of ATP and two molecules of guanosine triphosphate (GTP) (figure 8.5).

The synthesis of glucose from very simple compounds such as carbon dioxide and water occurs only in photosynthetic organisms (see chapter 7). Heterotrophs are able to synthesise glucose from relatively simple materials by a process referred to as **gluconeogenesis**. In most organisms this pathway requires compounds with at least three carbon atoms, such as pyruvate, to start the synthetic process. The ability to synthesise glucose is important because under fasting conditions a mammal will have enough glycogen in the liver to last for only about twelve hours or less.

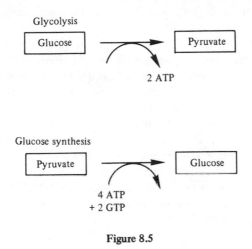

Figure 8.5

GLYCOGEN SYNTHESIS

Among the more simple anabolic processes following on from glucose synthesis is the formation of polysaccharide carbohydrates. The most common polysaccharides are glycogen, starch and cellulose. For our purposes here, glycogen synthesis will serve as an example to illustrate the basis of the process of polysaccharide formation.

Polysaccharide molecules are essentially a repeated series of simple units which form a chain. In the case of glycogen the basic unit is glucose (figure 8.6). In order to bring about synthesis of the polysaccharide chain, a series of repeated reactions takes place in which each glucose molecule is attached to the next; only a limited number of types of chemical linkage

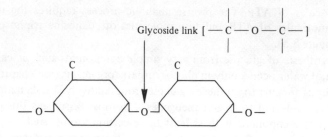

Figure 8.6 *Part of a glycogen molecule, showing two glucose units.*

are involved. In glycogen, each linkage is the same, being a glycoside linkage.

Here, another general feature of synthetic processes should be mentioned. The attachment of one simple unit to another frequently entails a whole series of chemical reactions in order to achieve a single linkage. Part of the reason for this is that it is essential that **activated** or 'high-energy' intermediates, derived from the basic building block are formed. These intermediates are capable of engaging in subsequent reactions which in themselves are exergonic. In order to achieve the formation of these 'activated' intermediates, ATP or other compounds with similar properties are involved such as uridine triphosphate in glycogen synthesis. In the synthesis of glycogen there are six reactions involved in the formation of each glycoside link between one glucose unit and another in the growing polysaccharide chain (figure 8.7).

FATTY ACIDS

Lipids consist of two types of building blocks; these are long-chain fatty acids linked by the other units which are glycerol. The significance of this group of compounds in biology is wide ranging and includes the storage of energy: they are important components of cell membranes and they are essential constituents of steroid hormones, such as the sex hormones and chemicals called prostaglandins. Synthesis of fatty acids occurs in the cytosol and oxidative catabolism of these compounds takes place in the mitochondria. Synthesis and catabolism each require a totally different set of enzymes.

Though lipids have smaller molecules than polysaccharides, their biosynthesis is more complex because more than one type of chemical linkage is involved. For our purposes we will consider only the synthesis of fatty acids here.

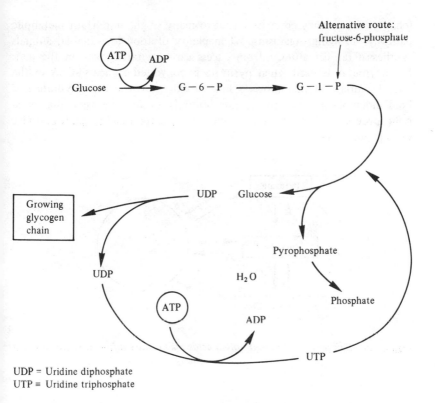

UDP = Uridine diphosphate
UTP = Uridine triphosphate

Figure 8.7 *Series of reactions in the formation of each glycoside link.*

As early as 1907 it was appreciated that fatty acids always contained an even number of carbon atoms. From this observation it was suggested that fatty acids were formed as the result of the condensation of a series of units each containing two carbon atoms. Later in the 1940s using deuterium (^{2}H) and ^{13}C, it was found that both of these isotopes were incorporated into fatty acids. Soon after this discovery the activated 2-C compound used in the synthesis of fatty acids was shown to be acetyl-CoA. This is a good example of early deductive work being supported by later findings made possible by improved techniques. Most of our information about fatty-acid synthesis comes from the study of the bacterium *Escherichia coli,* and although there are differences in the synthesis between different types of organisms, the fundamental principles appear to be the same.

The initial reaction of fatty-acid synthesis occurs in the cytosol and involves co-enzyme A. As will be appreciated from previous discussion (chapter 6), co-enzyme A is central to many of the important metabolic pathways in living organisms. When plenty of food is available, animals synthesise fat for storage from excess carbohydrates taken in the diet. Co-enzyme A is used when pyruvate is converted to acetyl-CoA in the mitochondria and this allows a link to exist between carbohydrate and lipid metabolism. Also, to increase interrelationships between the metabolic processes, acetyl-CoA is formed from certain amino acids and vice versa (figure 8.8).

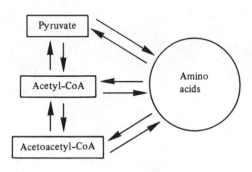

Figure 8.8 *Synthesis of different amino acids may involve different routes through the system.*

Acetyl-CoA is unable to pass through the mitochondrial membranes, therefore to overcome this problem, when it is formed from the breakdown of pyruvate, it is converted to citrate which is able to pass through the membrane; this is achieved by the addition of oxaloacetate. The citrate crosses the membrane and, when in the cytosol, it is converted back to acetyl-CoA and oxaloacetate, a process which needs to be coupled to the hydrolysis of ATP (figure 8.9).

Synthesis of fatty acids is brought about by a multi-enzyme complex called fatty acid synthase. The process involves the formation of activated building blocks as intermediates. In fact, although the basic building block was identified as a two-carbon compound many years ago, this view has been subsequently modified. Now, while a two-carbon compound is considered to be the first unit in the fatty-acid chain, following on from this, further two-carbon units are added by using three-carbon units (malonyl group) and removing one carbon atom as carbon dioxide when the linkage is made. Also the required reducing power is provided by reduced $NADP^+$. In order to produce the necessary malonyl groups for synthesis, the acetyl-CoA is converted to malonyl-CoA at the expense of the hydrolysis of

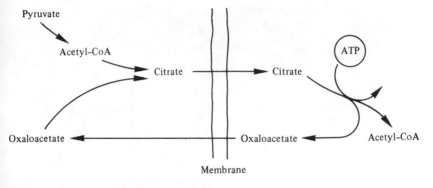

Figure 8.9

ATP. Malonyl-CoA therefore is the activated building block which is then added to the growing fatty-acid chain (figure 8.10).

Acetyl-CoA carboxylase, which is a controlling reaction for fatty-acid synthesis, may occur in either an active or an inactive form. The presence of citrate in adequate concentrations shifts the equilibrium for the enzyme from the inactive to the active form. In this way the presence of citrate regulates fatty-acid synthesis even though the citrate is not directly involved in the reaction.

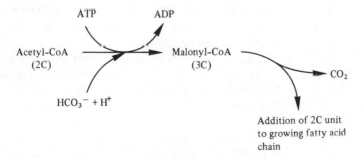

Figure 8.10

Many tissues, especially those involved in a good deal of lipid synthesis, have an alternative pathway for glucose breakdown in the cytosol apart from glycolysis. Here glucose-6-phosphate is dehydrogenated to 6-phosphogluconate. For this reason it is often referred to as the **phosphogluconate pathway**. One of the main purposes of this alternative sequence of reactions to glycolysis is to produce NADPH which is then available for reduction reactions as required. A supply of reduced NADP is needed for fatty acid synthesis in lipid formation (figure 8.11).

133

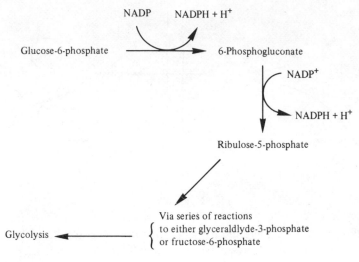

Figure 8.11

AMINO ACID SYNTHESIS

There are twenty naturally occurring amino acids; these, or their precursors, are produced from intermediate compounds in the respiratory pathway. For instance, glutamic acid is formed from α-ketoglutarate and this is then formed into glutamine and proline. Aspartic acid is formed from oxaloacetate, and alanine from pyruvate.

Generally the biosynthetic pathways are short. A typical simple route is the formation of glutamine which starts with the amination of α-ketoglutarate from the TCA cycle; this also includes a reductive reaction using NADH (or NADPH in plants).

$$NH_4^+ + \alpha\text{-ketoglutarate} + NADH \text{ (or NADPH)} + H^+$$
$$\rightarrow \text{glutamate} + NAD^+ \text{ (or NADP}^+) + H_2O$$

From here the amino group of glutamate is transferred to other organic acids to produce further amino acids.

The pathway outlined above occurs only in certain organisms when there are high concentrations of ammonia present.

A more frequently used method of incorporating free ammonia into organic compounds involves the two enzymes glutamine synthase and glutamate synthase (figure 8.12). This can occur at much lower concentrations of ammonia than those in the previous reaction. Both require ATP hydrolysis.

134

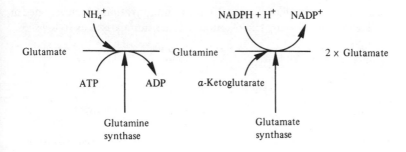

Figure 8.12

PROTEIN SYNTHESIS

The amino acids formed in the previous section may then be used as the building blocks for protein synthesis. Here, our concern is not with the way in which the genetic code of DNA interacts with the various types of RNA in the cell to form proteins, as this information can be found in other texts, but rather it is with some of the energetic considerations involved.

In protein synthesis the amino acids first need to be activated using ATP. This forms a complex with transfer RNA called **amino acyl-transfer RNA** (figure 8.13). The complex formed acts as both a carrier and specific adaptor for the triplet codons with a high-energy linkage between the amino acid and the carrier. Each amino acyl-transfer RNA complex is specific to a particular amino acid. The activation reaction occurs in the

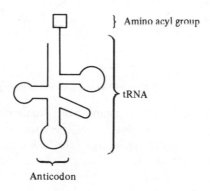

Figure 8.13 *Amino acyl-tRNA complex.*

135

soluble cytoplasm of the cell with 20 different activation enzymes, one for each different amino acid. The reaction may be summarised as follows

$$\text{amino acid* + ATP + tRNA} \xrightarrow[\text{enzyme*}]{\text{specific activating}}$$

amino acyl–tRNA* + AMP + pyrophosphate $\quad \Delta G = -27.2$ kJ

(*denotes a specific amino acid, tRNA or amino acyl–tRNA synthetase)

Biosynthesis of a polypeptide involves matching the activated tRNA to the appropriate complementary codon on the mRNA on the ribosome (figure 8.14). The sequence of matching and attaching an activated tRNA molecule with its specific amino acids so that it can be used in the build-up of a polypeptide chain occurs with the following main steps for each amino acid added. The activated amino acid–tRNA complex is first selected on to the relevant codon of the mRNA, and to achieve the required bonding a molecule of guanosine triphosphate (GTP) is used. A **peptide bond** is then formed between adjacent amino acids. The enzyme involved causes the transfer of the amino acyl group from the tRNA of one amino acid to the free amino group of the amino acyl–tRNA on the next site. As a result of this, the two amino acids are attached by the carboxyl group

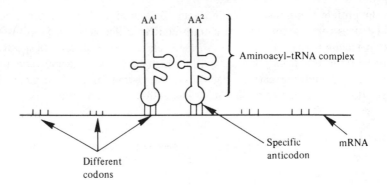

Figure 8.14

of one to the tRNA of the next amino acid. Neither ATP nor GTP is needed for formation of the peptide bond. The energetic requirements of this reaction are satisfied by the breaking of the high-energy bond between the amino acid and the tRNA.

BIOSYNTHESIS – SUMMARY

1. Entropy decreases when synthesis occurs – the reactions are therefore basically endergonic.

2. Such reactions are not spontaneous and must be coupled to reactions which are exergonic, such as the hydrolysis of ATP and other high-energy compounds.

3. Apart from ATP, other phosphorylated compounds may be involved, such as GTP, CTP or UTP, whose function is to transfer phosphate-bound energy from ATP to other specific biosynthetic pathways.

4. Many heterotrophs can produce cell components from glucose, amino acids, trace elements and a limited range of other materials in small amounts.

5. Catabolism and anabolism can share parts of the same sequence, but they cannot use complete common pathways because certain steps involve relatively large negative free-energy changes which are irreversible. Thus alternative portions to the pathway which use different enzymes must be used.

6. The formation of a single linkage in a synthetic process often involves a whole series of chemical reactions. In these reactions an activated intermediate is formed which takes part in subsequent exergonic reactions.

7. Heterotrophs can synthesise glucose by gluconeogenesis. This process can use pyruvate as a starting point.

8. In the synthesis of polysaccharides a series of repeated reactions takes place, often forming only one or two different types of linkage.

9. Lipids consist of long-chain fatty acids linked to glycerol.

10. Synthesis of fatty acids occurs in the cytosol and catabolism takes place in mitochondria, both processes using different enzyme systems.

11. Acetyl-CoA is the activated 2-C compound involved in fatty-acid synthesis.

12. Citrate is passed through the mitochondrial membranes to the cytosol where it is converted to acetyl-CoA and oxaloacetate; this is coupled to ATP hydrolysis.

13. Fatty acid synthesis is brought about by a multi-enzyme complex called fatty acid synthase.

14. The first building block used in fatty-acid synthesis is acetyl-CoA; subsequent units are added using malonyl groups (3-C) and removing CO_2.

15. The conversion of acetyl-CoA to malonyl-CoA is coupled to ATP hydrolysis.

16. Reduction reactions are also involved in fatty-acid synthesis. This is provided by NADP. Extra reduced NADP is formed in the cytosol by the phosphogluconate pathway.

17. In some organisms where high concentrations of ammonia are present, amino acid synthesis begins with glutamate formation by the amination of organic acids in the TCA cycle. This is followed by transfer of the amino group of glutamate to other organic acids from which further amino acids are produced.

18. More often, free ammonia is incorporated into organic compounds using glutamine synthase and glutamate synthase.

19. In protein synthesis the amino acids are activated by coupling ATP hydrolysis to the formation of acyl–transfer RNA complex.

20. The complex acts as a carrier and a specific adaptor for the triplet codons with a high-energy linkage between the amino acid and the carrier.

21. Synthesis of polypeptides involves matching the activated tRNA with the complementary codon on the mRNA.

22. Bonding between the tRNA complex and codon of the mRNA utilises a molecule of GTP.

23. A peptide bond is formed between adjacent amino acids, and here the amino acyl group from the tRNA of one amino acid is transferred to the free amino group of the amino acyl–tRNA on the next site.

9 Nerve Impulse Conduction

These last two chapters have been included to illustrate some principles relating to the energy relationships of certain familiar physiological functions, with particular emphasis on the way in which ATP may be used. Although the topics of these two chapters appear to be linked, the manner in which the energy relationships function is very different. Other examples could have been chosen but the number discussed had to be limited; the examples have to some extent been arbitrary, although these two topics are good examples of how biochemists, physiologists and those working on the fine structure of cells have co-operated to provide the type of overall picture that we have today.

RESTING POTENTIAL

Figure 9.1 shows the structure of part of a typical motor neurone; however, little knowledge of this structure is needed to understand the principles of function.

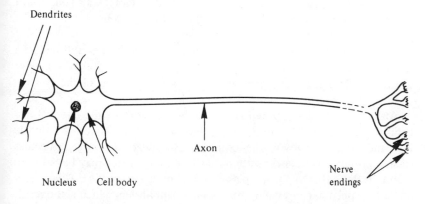

Figure 9.1 *Structure of a typical motor neurone.*

When the neurone is at rest, particular concentrations of ions, especially Na^+ and K^+, are maintained on either side of the axon membrane. Maintenance of these concentrations is achieved by the continual action of the Na^+-K^+-ATPase pump (see chapter 5). The result is that the concentrations of Na^+ and K^+ are different on either side of the membrane, with K^+ more concentrated on the inner surface and Na^+ on the outer surface (figure 9.2). This ionic imbalance causes a difference in electrical charge on either side of the membrane so that the inside is negative with respect to the

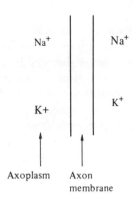

Figure 9.2

outside. This means that an electrical potential (that is, a voltage difference) exists across the membrane, which is referred to as the **resting potential**. The size of the electrical potential depends on the differences in the concentration of charged ions on either side of the membrane. Usually it is about 60 mV (between 50 and 70 mV). At this stage it is also important to point out that the membrane of the resting neurone is much more permeable to K^+ ions than to Na^+ ions.

The resting nerve is in dynamic equilibrium which is established as a result of the action of the Na^+-K^+-ATPase pump and the forces of passive diffusion across the membrane which operate as a result of the concentration differences in the ions on either side of the membrane (figure 9.3).

Although the resting potential is mainly the result of the unequal distribution of Na^+ and K^+ ions, it also depends on the balance of chloride ions and certain anions in the axoplasm, particularly for large molecules such as proteins. The concentrations of Cl^- on the outside may be between 5 and 10 times that of the axoplasm on the inside (figure 9.4). This contributes to a phenomenon called the **Donnan equilibrium** which is the result of passive processes but leads to the unequal distribution of ions. The

140

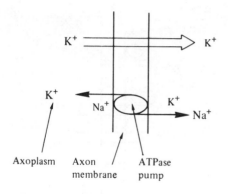

Axoplasm   Axon   ATPase
          membrane   pump

Figure 9.3

proteins inside the axon are large molecules which act as anions and at the same time they cannot pass through the axon membrane. K^+ is the counter (balancing) ion in the axoplasm. As we have seen, the concentration of Cl^- is greater outside the membrane than inside, consequently Cl^- diffuse inwards down their own concentration gradient. Since the membrane is relatively impermeable to Na^+ and permeable to K^+, the K^+ acts again as the balancing ion in order to maintain electroneutrality. The equilibrium concentration of K^+ inside the axon is therefore higher than Cl^-, and also the K^+ concentration is higher inside than outside. In this way a passive Donnan equilibrium is established with a higher concentration of K^+ ions inside than outside (figure 9.5). At the same time, although the passive forces of diffusion and the electrochemical gradient tend to move sodium into the axoplasm, the relative impermeability of the membrane to Na^+ and the action of the Na^+-K^+-ATPase pump maintain a high concentration of Na^+ outside. Thus the axoplasm of resting neurones

Ionic concentration (mM per Kg H_2O)

	Axoplasm	*Blood*
Na^+	50	440
K^+	400	20
Cl^-	120	560
Organic anions	360	—

Figure 9.4 *(after Hodgkin, A. L., The Conduction of the Nervous Impulse, Liverpool University Press, 1964).*

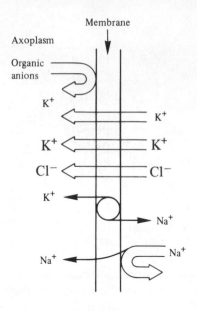

Figure 9.5

contains more K^+ and less Na^+ and Cl^- than is usual in other animal tissues or fluids.

The high K^+ concentration in the axoplasm means that there would normally be a tendency for these ions to move out across the membrane down a concentration gradient. However, there also exists an electrical gradient due to the electrical potential difference across the membrane, and this tends to hold the K^+ inside. Thus the forces of diffusion and electrostatic forces are in opposition in relation to K^+ (figure 9.6). The electrostatic force is normally of the order of -60 mV, as stated previously, while the diffusional forces as a result of the concentration gradient have been estimated at about $+75$ mV. This means that the electrostatic force is more than cancelled by the diffusional force, giving an overall driving force of 15 mV to move K^+ out of the axoplasm (figure 9.6).

Investigations using radioactively labelled K^+ indicate that potassium ions are being continually lost from the axoplasm across the membrane of a resting neurone. The net loss is compensated by the Na^+-K^+-ATPase pump. Metabolic inhibitors, such as dinitrophenol (DNP) prevent the synthesis of ATP, therefore the pump ceases to function but this does not affect passive processes. Experiments with DNP reduce the inward movement of K^+ but have no effect on the outward movement of the same ions. This is evidence for the active pump mechanism returning K^+ to the axoplasm (figure 9.7).

142

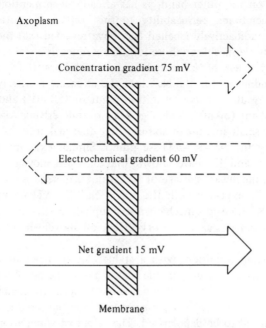

Axoplasm

Concentration gradient 75 mV

Electrochemical gradient 60 mV

Net gradient 15 mV

Membrane

Figure 9.6

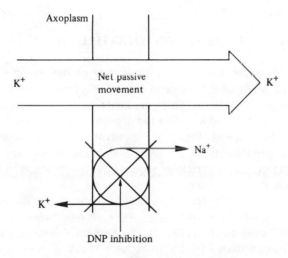

Axoplasm

K^+

Net passive movement

K^+

Na^+

K^+

DNP inhibition

Figure 9.7

With Na$^+$ on the other hand, as has already been mentioned, the difference in membrane permeability to this ion has a profound effect. Studies using radioactively labelled Na$^+$ have revealed that the membrane is only slightly permeable to those ions. Consequently there is a very small amount of movement of Na$^+$ into the axoplasm by passive forces, although the forces tending to pull the Na$^+$ into the axon are relatively large since they are the result of a concentration gradient (50 mV) and an electrochemical gradient (60 mV) which give an overall driving force of about 110 mV. The small amount of sodium that does leak into the axoplasm is moved out by the Na$^+$-K$^+$-ATPase pump. Similar experiments to those involving DNP and K$^+$ have shown that outward movement of Na$^+$ is inhibited, but the inward passage of Na$^+$ is not affected.

As might be expected with the active Na$^+$-K$^+$-ATPase pump, where the Na$^+$ and K$^+$ transport mechanism is coupled, reduction of the external K$^+$ concentration has an inhibitory effect on movement of Na$^+$ out of the axoplasm.

In this resting condition, with a stable electrical potential difference between the inside and the outside of the axon, the membrane separates positive and negative charges. When it is in this state, the membrane is said to be **polarised**. However, if the membrane potential is reduced, the membrane is then said to be **depolarised**. The effect of stimulation is to cause this depolarisation of the membrane.

DEVELOPMENT OF THE ACTION POTENTIAL

The ionic differences produced on either side of the resting axon membrane are the result of ATP-dependent active transport processes which create the required gradients. These gradients are sufficient to cause rapid back-diffusion of Na$^+$ and K$^+$ when the neurone is stimulated.

Work by Hodgkin and Huxley has shown that very short-lived changes occur in the permeability of the axon membrane to Na$^+$ and K$^+$ when depolarisation occurs. A particular sequence of events seems to take place following stimulation of a neurone.

As we have seen, the resting potential of the membrane is at about -60 mV on the inside relative to the outside. Depolarisation takes place in about 1 millisecond and here, the electrical potential across the membrane is rapidly changed from -60 mV to about $+40$ mV. This being the **action potential**, this event is followed by a somewhat slower return to the resting potential. During this latter phase the electrical potential falls very briefly to a level below -60 mV, that is, about -75 mV (figure 9.8).

144

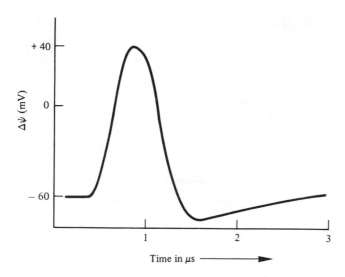

Figure 9.8 *Change in membrane potential at one point on the axon membrane when an action potential is generated.*

These events are related to very short-lived changes in the permeability of the axon membrane to Na^+ relative to K^+. The details of the mechanism for this at the membrane molecular level are not known, but are the subject of considerable speculation. However, the rapid rise in the electrical potential of the inside of the membrane to about +40 mV is accompanied by a considerable increase in the permeability to Na^+ and a decrease in permeability to K^+. Return to the resting potential is related to a great reduction in Na^+ permeability and an increase in K^+ permeability (figure 9.9). Change in the relative movement of ions through the membrane is referred to as the change in conductance of the membrane to the ions. This is measured in 'reciprocal ohms' (called mhos) per unit area of axon membrane.

Thus the electrical potential across the membrane determines the membrane permeability to Na^+ and K^+. The resting potential is to a large extent due to the K^+ gradient since the unstimulated nerve membrane has a high permeability to this cation only. When depolarisation takes place, the membrane becomes much more permeable to Na^+ than to K^+ and, because of the ion gradient maintained in the resting state by the Na^+-K^+-ATPase pump, the Na^+ then moves rapidly into the axoplasm down its own electrochemical gradient. The resulting change in the electrical potential across the membrane caused by the increased Na^+ permeability has the effect of increasing permeability to K^+, which in turn moves down its own electrochemical gradient. This then means that there is a return to a

145

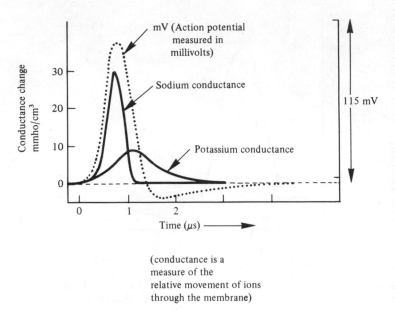

(conductance is a
measure of the
relative movement of ions
through the membrane)

Figure 9.9

negative electrical potential on the inside of the membrane relative to the outside, and this is accompanied by a reduction of the inward movement of Na^+. As the electrical potential across the membrane approaches the resting state, the K^+ and Na^+ permeabilities also return to their resting state (figure 9.10).

It is important to realise that actually very small changes in the proportions of K^+ and Na^+ cause relatively large changes in the electrical potential across the membrane; consequently these changes can take place extremely rapidly.

According to the Hodgkin and Huxley hypothesis, the events described and shown in figures 9.8, 9.9 and 9.10 occur at one point of the membrane of the axon at any particular time as an action potential is conducted along the membrane. The conduction takes the form of a wave of depolarisation which moves at a rate of up to about 20 metres per second along the membrane, followed by a return to the polarised state. When depolarisation takes place at a particular point on the membrane, this causes an electric current to flow, as shown in figure 9.11. Current flows from the relatively positive region on the inner surface of the membrane which is depolarised to the adjacent region which is relatively negative. This in turn has the effect of causing depolarisation and an action potential across the

146

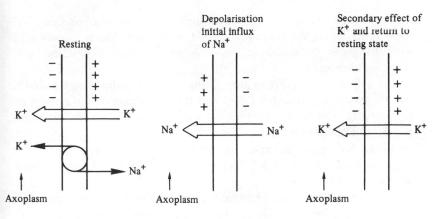

Figure 9.10

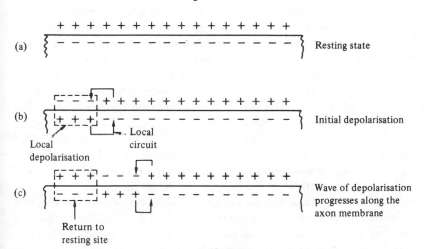

Figure 9.11

next adjacent region and so on; thus it is propagated along the membrane as an impulse.

By the mechanism described the nerve impulse is a phenomenon which reinforces itself as it moves along using the energy from ion gradients built up while the nerve has been in a resting state. If this self-reinforcement did not occur, the electrical current propagated along the axon membrane would rapidly decrease over a short distance as a result of the effects of leakage and resistance.

The nerve impulse that passes along the axon membrane is caused by a reversal of the electrical potential across the membrane which travels along the membrane. These changes cause local circuit currents to flow, and these bring about depolarisation of the membrane, which is associated with changes in the membrane permeability which moves systematically along the membrane.

The action of the nerve in propagating a nerve impulse along its axon is due to passive movements of Na^+ and K^+ which result from membrane permeability. The return to the resting state is brought about by the active Na^+-K^+-ATPase pump.

As the Na^+ moves in rapidly, when depolarisation takes place there is some outward leakage of K^+. This has been demonstrated by the use of radioactively labelled K^+.

Part of the Hodgkin and Huxley hypothesis is that changes in the membrane permeability are due to the passive movement of K^+ and Na^+ through separate channels in the membrane. This has been subsequently supported by experiments involving specific blocking agents for either, but not both, of the ions. The current theory is that these passive ionic movements are through separate channels which are controlled chemically in a specific way. These are often referred to as gated channels. Thus, when the membrane is depolarised by more than 15 mV, which appears to be a threshold level for generation of an action potential, the gates in the Na^+ channel open to some extent so that Na^+ gain exceeds K^+ loss. The greater the depolarisation above 15 mV, the more the Na^+ channel opens, and this allows Na^+ movement in one direction to even further exceed K^+ movement in the other. At levels above the critical level of 15 mV the Na^+ ions move into the axon along concentration and electrochemical gradients, and this causes further local depolarisation in the membrane. This is also self-reinforcing so that the inward movement of Na^+ causes the channels to open further to allow the steep rapid inrush of Na^+, and the consequent rapid reversal of charge which occurs locally at the membrane surface. Then, at a critical point, the permeability to Na^+ changes and it becomes more permeable to K^+. The K^+ gradient is great and this causes rapid movement of K^+ out of the axon. Thus, the charge of the outgoing K^+ repolarises the membrane. The membrane is returned to the resting potential by the small quantities of K^+ and Na^+ which are then exchanged by the Na^+-K^+-ATPase pump (figure 9.12).

THE VOLTAGE-CLAMP TECHNIQUE

A significant aid to investigation of nerve membrane physiology was the discovery of a method known as the **voltage-clamp technique**. This was developed by Cole in 1949 and was later used in the important investi-

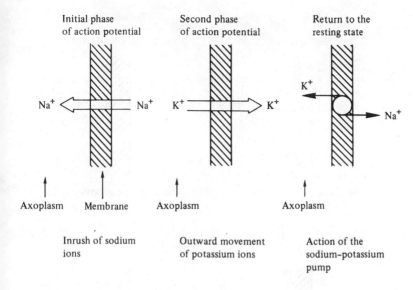

Initial phase of action potential

Second phase of action potential

Return to the resting state

Axoplasm   Membrane   Axoplasm   Axoplasm

Inrush of sodium ions

Outward movement of potassium ions

Action of the sodium–potassium pump

Figure 9.12

gations of Hodgkin and Huxley. The use of this method enabled researchers to adjust the membrane potential to any desired level and then to hold this steady. In the resulting voltage-clamped state, ionic movements can be estimated by finding the current required to hold the membrane potential at the required level. This provided workers with a technique with which to investigate selectively the effects of membrane electrical potential on the movement of selected ions.

Part of the procedure involved depolarising the membrane to a degree which would bring about the normal permeability changes associated with conduction of a nerve impulse. In these experiments, the changes in electrical charge are compensated by an electrical device called a feedback amplifier so that the membrane potential remains unchanged. In this state, the permeability changes which follow depolarisation occur in a somewhat modified form but there is no action potential; that is to say, there is no impulse (figure 9.13). Here it should be borne in mind that when investigatory methods are used which appear to modify the natural state, then caution is needed on the part of research workers when interpreting their findings. Frequently checking results obtained when using different methods, and comparing these one with another, is essential.

The voltage-clamp method can also be used to help work out the sequence of changes in conductance which occur when an axon membrane conducts an impulse. The relationship between membrane potential and

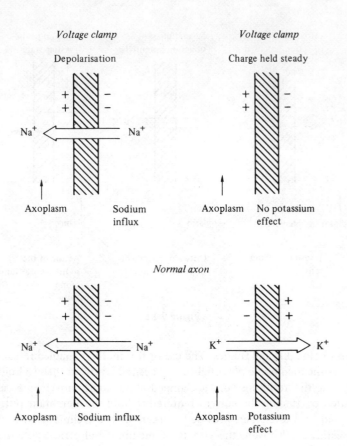

Figure 9.13

ionic permeability is important in the generation of a nerve impulse. However, this technique does not distinguish between the different types of ions being passed through the membrane, and thus either K^+ or Na^+ may be concerned.

Hodgkin and Huxley were able to deduce that an action potential is accompanied by movements of Na^+ and K^+ and then, with the aid of voltage-clamp experiments, further details were hypothesised.

It had been found that membrane currents (measured in μA per unit area of axon membrane surface) during an action potential consist of two phases, with a brief inward current followed by a longer outward current (figure 9.14). Use of voltage-clamp preparations in solutions with low sodium concentrations outside the membrane showed that the early phase of the current, because of a surge of Na^+ through the membrane, was

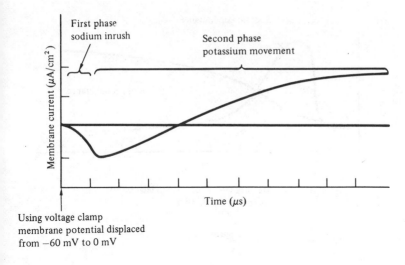

Figure 9.14

always reduced exactly in proportion to any reductions made in the external Na^+ concentration. Also, it was found that the current that occurs after the initial surge of Na^+ is unaffected by changes in the Na^+ concentration outside the membrane. Hodgkin and Huxley found that changes in membrane permeability, which arise when depolarisation is achieved with voltage clamps, take time to decay when the clamp is removed. Thus the permeability changes established by the clamp during depolarisation continue briefly after membrane repolarisation and current continues to flow. The assumption made here was that the delayed current is also caused by the same factors as the final part of the change in potential that accompanies an action potential. This delayed current was outward with respect to the resting potential, and it could be reversed when the membrane potential exceeded the resting potential by more than 12 mV. These findings provide further evidence, since this delayed current has a value that is about equal to the resting potential (60 mV) plus 12 mV (that is, 72 mV), which is of the order of that estimated for the movement of potassium in the normal axon.

This supports the view that the first part of the action potential is due to movement of Na^+ and the last part is due to movement of K^+ (figure 9.15).

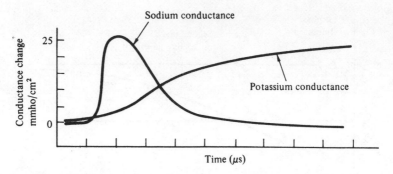

Membrane conductance in relation to
sodium and potassium

Figure 9.15

NERVE IMPULSE CONDUCTION – SUMMARY

1. In the resting state the axon membrane has a higher concentration of K^+ on the inside and Na^+ on the outside. The difference is maintained by the Na^+-K^+-ATPase pump.
2. Ionic imbalance causes the inside of the membrane to be more negative with respect to the outside – thus there is an electrical potential across the membrane.
3. The resting membrane is more permeable to K^+ than Na^+.
4. Chloride ions and charged protein molecules in the axoplasm also contribute to the resting potential. A Donnan equilibrium occurs which also leads to unequal distribution of ions on either side of the membrane.
5. K^+ ions act as the balancing ion in the Donnan equilibrium; this also leads to greater imbalance in the concentration of K^+ on either side of the membrane.
6. The tendency for K^+ ions to move out of the axoplasm is reduced by the electrical gradient across the membrane.
7. The relatively small net loss of K^+ that does occur is compensated by the Na^+-K^+-ATPase pump. Here K^+ is exchanged for the small amount of Na^+ that has leaked in.
8. The effect of stimulation of an axon is to depolarise the membrane.
9. When the axon is stimulated, transitory changes occur in the permeability of the membrane to Na^+ and K^+, so that Na^+ passes through much more readily and K^+ less easily.

10. On depolarisation the electrical potential across the membrane changes rapidly from -60 mV to $+40$ mV. This is followed by a slower return to the resting potential.

11. Return to the resting potential is related to great reduction in Na^+ permeability and increase in K^+ permeability.

12. Electrical potential across the membrane determines membrane permeability to Na^+ and K^+.

13. At depolarisation the Na^+ move rapidly into the axoplasm.

14. These changes have the effect of increasing permeability to K^+ which then moves down its electrochemical gradient and this returns the inside to a negative electrical potential relative to the outside.

15. As the electrical potential across the membrane approaches the resting state, the permeability to K^+ and Na^+ returns to that of the resting state.

16. Small changes in the proportions of K^+ and Na^+ on either side of the membrane cause relatively large changes in electrical potential – thus changes can take place rapidly.

17. Conduction of a nerve impulse takes the form of a wave of depolarisation which moves along the membrane, followed by a return to the polarised state.

18. Changes in electrical potential across the membrane as a result of depolarisation cause local circuit currents to flow and these bring about depolarisation in adjacent regions of the membrane which passes systematically along the membrane.

19. Return to the resting state is also brought about by the Na^+-K^+-ATPase pump.

20. Recent theory is that Na^+ and K^+ move through the membrane via separate gated channels which respond to changes in the electrical potential across the membrane.

21. Use of the voltage-clamp technique, which enables researchers to adjust membrane potential to any desired level, has been important in elucidating the mechanisms involved.

22. When a voltage clamp is used, permeability changes occur, but there is no action potential.

23. Using the voltage-clamp method has shown that the early phase of the action potential was due to a surge of Na^+ and the later phase was due to outward movement of K^+.

10   Muscle Contraction

The final area relating to the application of cellular energetics in living organisms is discussion of the phenomenon of muscular contraction. This has been chosen because it is a familiar example of an energy-conversion process whereby chemical energy (ATP) is converted into mechanical energy. Muscle is therefore an energy-converting device. One of the problems, however, is that although there have been considerable developments in the study of this area of muscle cellular physiology, there are still many aspects that are unknown or remain very speculative. Therefore it will be a combination of this empirical or observational evidence, together with more speculative ideas, which will be under consideration here.

Contractile elements are not uncommon in living things; for instance, there are cilia and flagella (hair-like structures of microscopic size, which many micro-organisms, and some tissues of higher organisms, possess). The spindles involved in cell division have a contractile function, as have the cytoplasmic fibres which take part in cell differentiation. However, the best understood are the contractile properties of muscle, particularly skeletal muscle; it is with skeletal muscle that we are concerned in this chapter.

An important contributory factor to the conversion of the chemical energy of ATP into mechanical energy is the very special structure of muscle and the way in which this is linked to the biochemistry involved. Consequently, although we have not been particularly concerned with structure in our previous discussion (although it is mentioned to some extent with reference to enzymes), muscle is a good example of the relationship between structure and function. Indeed it is essential to appreciate something of the structure of muscle in order to understand the nature of the energy conversion that takes place. Here, we are especially helped by the very detailed work carried out by electron microscopists in relation to the detailed structure of muscular tissue.

THE STRUCTURE OF SKELETAL MUSCLE

Skeletal muscle consists of very long multi-nucleated cells. Each of these cells contains contractile elements called **myofibrils**, which are also very elongated. The myofibrils are arranged in bundles so that they are parallel to each other along their long axis. Each myofibril is made up of **myofilaments** which are elongated and run parallel to each other. Thus, each muscle cell has a bundle of myofibrils and each of these has a bundle of myofilaments (figure 10.1).

If we now consider one myofibril it can be seen that it has a repeated pattern of light and dark bands along its length. These can be divided into units called **sarcomeres** and these are the individual contractile portions of the myofibrils.

Great detail of the structure is not necessary but sufficient needs to be known to be able to relate structure to function. High-magnification electron microscopy has been of considerable assistance in the study of muscle function, particularly in work by Huxley and colleagues. These workers were able to show that myofibrils contain two types of filament, one thick and the other thin, arranged in a repeated pattern (figure 10.1). From the thick filaments there are regularly placed projections which form crosslinks with the thin filaments (figure 10.2). When the myofibril contracts, the length of both thick and thin filaments remains constant. Changes in

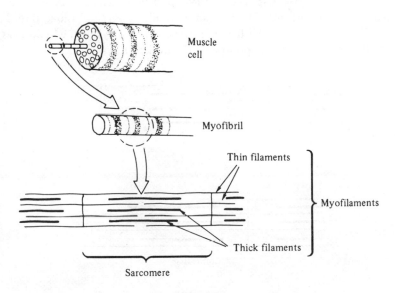

Muscle cell

Myofibril

Thin filaments

Myofilaments

Thick filaments

Sarcomere

Figure 10.1

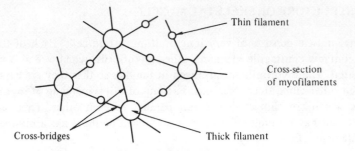

Thin filament

Cross-section
of myofilaments

Cross-bridges

Thick filament

Figure 10.2

myofibril length, and therefore muscle length, are due to sliding of thick and thin filaments along each other (figure 10.3). In order that this can take place, the cross-links between the filaments must be broken and reformed elsewhere along their length.

CHEMICAL COMPOSITION OF MUSCLE

One of the main chemical constituents of muscle is the protein, **myosin**; this is made up of extremely long polypeptide chains (see chapter 8). Myosin extracts have also been shown to behave as an enzyme when calcium ions are present. The myosin molecule consists of a long thread-like tail with a 'head' which has ATPase activity catalysing the hydrolysis of ATP to ADP and organic phosphate (figure 10.4).

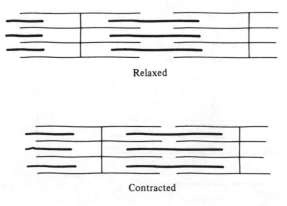

Relaxed

Contracted

Figure 10.3

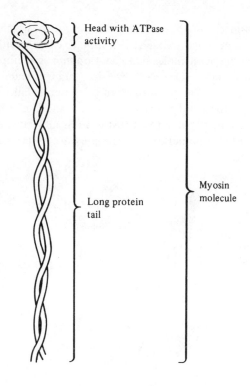

Figure 10.4

In the early 1940s Szent-Györgyi, the Hungarian biologist, discovered a second protein in muscle, which is **actin**. Together actin and myosin make up over 75 per cent of the protein of the myofibrils. Actin is within the thin fibres and myosin is in the thick fibres. Purified extracts of these two proteins bind together to form a complex called **actomyosin** and this combination, which takes place outside the cell, appears to have an important consequence for the contraction mechanism of muscle inside the cell.

THEORIES OF MUSCULAR CONTRACTION

The Huxley hyothesis uses the idea that the thick filaments consist of myosin, with the ATPase part being within that part of the molecule which forms the cross-link with the thin filament. However, the thin filament consists not only of actin, which is a long double-chain polypeptide protein, but also two other proteins, **tropomyosin** and **troponin**.

157

The tropomyosin is also a very long-chain protein which fits into the groove of the double actin strand; it is not fixed in position but it can move along on the actin molecule. The troponin is in three parts, each with a specific function. One part reversibly binds calcium ions, another binds to actin and can inhibit formation of the cross-links between actin and myosin, and the third binds to tropomyosin. The troponin is fixed to the tropomyosin, but not fixed to the actin molecule, its attachment depending on it being bound only in the presence of calcium ions (figure 10.5).

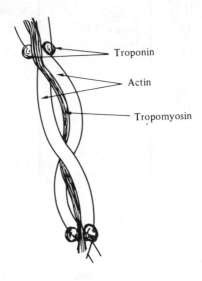

Troponin

Actin

Tropomyosin

Figure 10.5 *Protein arrangement in a thin filament.*

As has been previously mentioned, mixtures of actin and myosin form the complex actomyosin. Evidence from electron microscopy confirms that the portion of the myosin molecule which has the ATPase activity binds to the actin. The complex dissociates when ATP and magnesium ions are present. Following this dissociation, ATP hydrolysis occurs and after this the complex is reformed. The action of formation and dissociation of the complex involves the making and breaking of cross-links between the thick and thin filaments.

Within skeletal muscle, magnesium and calcium ion concentration are important. When the muscle is in a relaxed state there is a low concentration of calcium ions. In this condition there are few cross-links. The portion of the myosin molecule with ATPase activity is able to bind two molecules of either ATP or ADP and inorganic phosphate (figure 10.6).

In the relaxed state, the myosin is unable to react with the actin filament because of lack of calcium ions in the medium. Motor nerve impulses have the effect of releasing calcium ions into the medium surrounding the myofibrils. The calcium then binds to the troponin molecules on the thin filament. The effect of this is to bring about a change in form of the

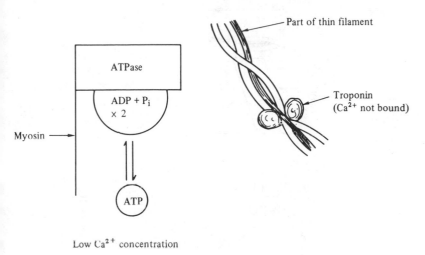

Figure 10.6 *Muscle in the relaxed state.*

troponin molecule, which in turn causes the myosin ATPase portion of the myosin molecule to attach to the actin. When this occurs, a change in form takes place in the myosin ATPase region, which causes an alteration in its angular position in relation to the filamentous part of the myosin molecule. This then makes the thin filament slide along the thick filament. At the same time the ATPase portion of the myosin molecule changes its structure and the ADP and inorganic phosphate dissociate from this part of the molecule. Two ATP molecules are hydrolysed for each cross-bridge broken. It should be appreciated then that ATP hydrolysis is required to break cross-links and not to form them (figure 10.7). Breakdown of the bound ATP to produce bound ADP and P_i results in the myosin ATPase head moving back to its original angle so that it can attach to the next part of the thin filament, thus drawing it along.

Exactly how the ATP energy is used for producing the movement is not known. However, an additional hypothesis is that when myosin and

159

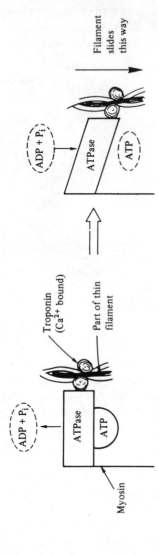

Myosin

Troponin
(Ca²⁺ bound)

Part of thin
filament

ADP + P$_i$

ATPase

ATP

ADP + P$_i$

ATPase

ATP

Filament
slides
this way

The ATPase part of the myosin attaches to troponin,
altering the angular alignment of the 'head' of the myosin
molecule with the rest of the molecule.

Figure 10.7

ATP react, hydrogen ions are formed rapidly and this leaves an **ADP-phos-phate-myosin** complex where the free-energy change of hydrolysis is conserved in the new conformation of the molecule. The complex then changes more slowly to yet another complex which also contains ADP-phosphate and myosin but it is in a 'low-energy' state; that is, it has a different conformation. The energy change here is in some way linked to the movement of the ATPase part of the myosin molecule, causing it to slide the actin filament along. A good deal of this part of the theory is very speculative.

INITIATION OF MUSCULAR CONTRACTION

Contraction of the myofibrils is triggered by a motor nerve impulse. Resting muscle has an electrical potential across the surrounding membrane of about 60 mV in a very similar way to that of a nerve axon (see chapter 9). Excitation causes depolarisation which leads to a rapid change in the permeability of the membrane to ions of sodium, potassium and calcium. In resting muscle, calcium ions are separated from the fluid surrounding the myofilaments but on stimulation the calcium rapidly moves in. ATP hydrolysis by myosin is very slow in resting muscles because of the inhi-bition by magnesium ions present, but this inhibition is overcome in the presence of calcium ions. When stimulation ceases, the calcium ions are moved out by the ATP-dependent calcium-ion pump (figure 10.8). Thus ATP hydrolysis is required for both contraction and relaxation of muscle.

PHOSPHOCREATINE AND MUSCLE FUNCTION

Muscle tissue contains the substance **phosphocreatine** in a concentration which is about five times that of the concentration of ATP. Phospho-creatine is a so-called 'high-energy' phosphate compound ($\Delta G^{\ominus} = -42$ kJ mol^{-1} for hydrolysis to creatine and inorganic phosphate). Under the conditions that exist in muscle tissues, the action of the enzyme creatine kinase causes the phosphate group of phosphocreatine to be readily trans-ferred to ADP to form ATP. Phosphocreatine has a much more negative standard free energy of hydrolysis than ATP, which also means that phosphocreatine more readily donates its phosphate to ATP

$$\text{phosphocreatine} + \text{ADP} \rightleftharpoons \text{creatine} + \text{ATP}$$

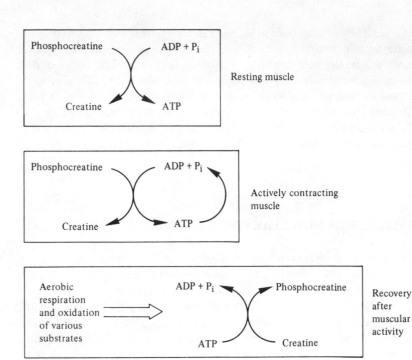

Figure 10.8 *Action of phosphocreatine in muscle.*

Although this reaction is reversible with the concentrations that occur in muscle tissue, the equilibrium is very much in the direction of ATP formation. Consequently, with the normal level of phosphocreatine in resting muscles, almost all of the ADP is phosphorylated to ATP.

Early investigations by Lindsgaard in 1931 showed that when iodo-acetate was used to inhibit glycolysis, or cyanide used to inhibit aerobic respiration, isolated muscle tissue will continue to contract when stimulated. The significance of this is that ATP formation has not occurred, yet muscle contraction takes place. His experiments showed no obvious utilisation of fuel materials but there was a reduction in the phospho-creatinine level with no reduction in the ATP level.

In muscle treated with creatine kinase inhibitor, then the phospho-creatine does not disappear when the muscle contracts, but the ATP level does decrease and ADP is formed. This also provides evidence that ATP, rather than phosphocreatine, is the immediate energy source for muscle contraction.

162

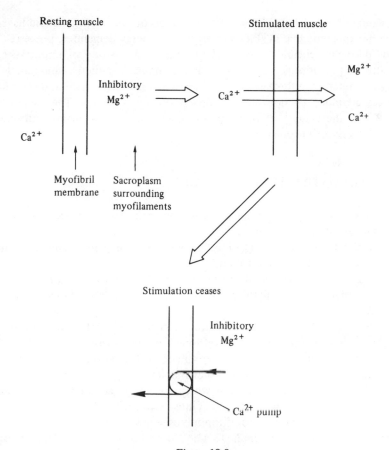

Figure 10.9

The suggested explanation of all these investigations is that ATP concentration is maintained at a high level in muscle tissue by the phosphocreatine present. The phosphocreatine is then regenerated during recovery periods in the muscle (figure 10.9). Here, when phosphocreatine levels are low after muscular activity, synthesis of phosphocreatine occurs at the expense of ATP formed during aerobic respiration and oxidation of various substrates in the resting muscle. In this way phosphocreatine is regenerated to its former level before muscular activity took place. Phosphocreatine, in these high concentrations in muscle, is able to act as a reserve to donate phosphate to ADP, as it is formed from ATP breakdown in actively contracting muscle. In invertebrates, arginine rather than creatine forms the basis of reserves in the muscle tissue.

163

From what has been discussed in the pages of this book, it should have become apparent that ATP is central to most energy-demanding processes. This particular chemical has an almost universal position of importance in all living organisms. Biosynthesis, nerve impulse conduction and muscle contraction are but three examples of the way that ATP is used in different energy-coupling reactions and energy-conversion processes. The study of ATP and the energy relationships of cells is one of the major unifying principles of cell biochemistry.

MUSCLE CONTRACTION – SUMMARY

1. Muscle contraction is an energy-conversion process where chemical energy (ATP) is converted to mechanical energy.
2. Skeletal muscle consists of cells made up of myofibrils and these in turn are made of myofilaments.
3. The individual contractile part of the myofibril is a sarcomere.
4. Myofilaments contain thick and thin filaments with cross-links between them.
5. When muscle contracts, the filaments slide along each other with the cross-links being broken, and reformed, in the process.
6. The thick filaments contain myosin which also has ATPase properties. This enzyme section is within the cross-link.
7. Thin filaments contain actin which forms the complex actomyosin when it combines with myosin in muscle contraction.
8. Thin filaments also contain tropomyosin and troponin.
9. Troponin reversibly binds to Ca^{2+}; attachment of the troponin to actin depends on binding in the presence of Ca^{2+}.
10. When actomyosin dissociates to form actin and myosin, ATP hydrolysis occurs; the complex then readily reforms again.
11. Thus formation and breaking of cross-links involve formation and dissociation of actomyosin which also involve ATP hydrolysis.
12. The presence of Ca^{2+} aids formation of actomyosin and in the relaxed state, the concentration of Ca^{2+} is below the threshold for this to occur.
13. Motor nerve impulses cause the electrical potential of muscle membranes to alter, thus changing the permeability of the membranes to Ca^{2+} which then move into the myofibrils.
14. When stimulation ceases, calcium ions are moved out by the ATP-dependent calcium pump. Thus ATP hydrolysis is needed for both contraction and relaxation of muscle.
15. Calcium binds to troponin in the thin filaments.

16. This causes the ATPase of myosin to attach to actin.
17. Changes in the angular position of the myosin ATPase unit in relation to the rest of the myosin molecule also occur, and this makes the thin filament slide along the thick one.
18. Also the ATPase changes its structure and the ADP and organic phosphate dissociate from the molecule.
19. Phosphocreatine occurs in large amounts in muscle tissue. Hydrolysis of phosphocreatine has $\Delta G^{\ominus} = -42$ kJ mol^{-1}.
20. Creatine kinase causes the phosphate of phosphocreatine to be transferred to ADP to form ATP.
21. Almost all ADP in resting muscle is converted to ATP by phosphocreatine. This keeps the ATP level high. Phosphocreatine is regenerated during recovery periods in muscle.

Bibliography and Suggestions for Further Reading

Anderson, J. W., *Bioenergetics of Autotrophs and Heterotrophs*, Edward Arnold, London, 1980.

Bagshaw, C. R., *Muscle Contraction*, Chapman and Hall, London and New York, 1982.

Chappell, J. B., *ATP*, Carolina Biological Supply Company, North Carolina, 1977.

Clayton, R. K., *Photosynthesis: Physical Mechanisms and Chemical Patterns*, Cambridge University Press, 1980.

Davies, M., *Functions of Biological Membranes*, Chapman and Hall, London and New York, 1973.

Dawber, J. G. and Moore, A. T., *Chemistry for the Life Sciences*, Macmillan, London, 2nd edn, 1980.

Edwards, N. A. and Hassall, K. A., *Biochemistry and Physiology of the Cell*, McGraw-Hill, New York and London, 2nd edn, 1980.

Jones, C. W., *Biological Energy Conservation: Oxidative Phosphorylation*, Chapman and Hall, London and New York, 2nd edn, 1981.

Lehninger, A. L., *Biochemistry*, Worth, New York, 2nd edn, 1975.

Lehninger, A. L., *Bioenergetics*, Benjamin Cummings, California and England, 2nd edn, 1973.

Morris, J. G., *A Biologist's Physical Chemistry*, Edward Arnold, London, 1974.

Tribe, M. and Whittaker, P., *Chloroplasts and Mitochondria*, Edward Arnold, London, 2nd edn, 1981.

West, I. C., *The Biochemistry of Membrane Transport*, Chapman and Hall, London and New York, 1983.

Wilkie, D. R., *Muscle*, Edward Arnold, London, 2nd edn, 1976.

Wood, E. J. and Pickering, W. R., *Introducing Biochemistry*, John Murray, London, 1982.

Zubay, G., *Biochemistry*, Addison-Wesley, London, 1983.

Index